Element of Risk
The Politics of Radon

Other books by Leonard A. Cole

Blacks in Power: A Comparative Study of Black and White Elected Officials
(Princeton University Press, 1976)

Politics and the Restraint of Science (Rowman and Allanheld, 1983)

Clouds of Secrecy: The Army's Germ Warfare Tests Over Populated Areas
(Rowman and Littlefield, 1988)

Element of Risk: The Politics of Radon
*was designed and illustrated by
Julie A. Cherry, and composed at
AAAS Press, a division of the
American Association for the
Advancement of Science,
on a Macintosh system using
New Caledonia, Modern 735 and
Folio type faces.
Printed by Quebecor Books,
Book Press, Inc.*

Element of Risk
The Politics of Radon

Leonard A. Cole

AAAS PRESS

A publishing division of the American Association
for the Advancement of Science

Library of Congress Cataloging-in-Publication Data

Cole, Leonard A., 1933–
    Element of risk : the politics of radon / Leonard A. Cole
        p. cm.
    Includes bibliographical references and index.
    ISBN 0-87168-513-2
    1. Radon—Environmental aspects.   2. Radon—Government policy—
United States.   I. Title.   II. Series.
Q181.A1A68   1993
[TD885.5.R33]
500 s—dc20
[363.17'99]   93-7272
                                                            CIP

AAAS Publication: 93-05H
International Standard Book Number: 0-87168-513-2
Printed in the United States of America on acid-free paper

TABLE OF **Contents**

Preface

I FIRST BECAME INTERESTED IN RADON ISSUES BECAUSE of a human interest story in northern New Jersey, the part of the state in which I live. Several houses in the Montclair area were found to have elevated radon levels, and in 1984 state officials announced plans to remove the surrounding soil from which the radioactive gas was seeping. The following year, upon urging by the officials, some of the homeowners moved out so that the soil could be excavated and replaced. The residents had been assured they would be able to return in a few weeks. As recounted in this book, however, a series of miscalculations by the authorities led to the forced abandonment of the homes and to unresolved anguish for the displaced families.

The source of the Montclair radon problem apparently was man-made. Soil from a defunct plant where radium was painted on watch dials presumably had been dumped there 60 years earlier. State and federal officials thought that digging up and removing the soil was the best way to mitigate radon concentrations in the houses above. But their inability to dispose of the excavated dirt caused continuing rancor among the public.

Around the same time, federal authorities announced suspicions that residential radon from naturally occurring sources posed a nationwide hazard. Coincidentally, geological conditions in northern New Jersey suggested that homes there might be especially prone to elevated radon concentrations. The finding was perplexing. Would environmental officials who were floundering over policy on radon from a man-made source be more competent in dealing with radon from a natural source?

While newspaper reporting about radon in the state was extensive, it left me uncertain about whether government officials were overstating, or perhaps understating, the problem. Moreover, it reached to the larger matter of what constituted proper government response to environmental and health risks.

My interest in these questions became the root of this inquiry. Beneath the surface of an apparently straightforward national radon policy, I learned that officials of one federal agency differed with those of another; respected scientists held disparate views; elected officials, environmental activists, radon testers and remediators, realtors, and affected citizens held strong though dissimilar positions. Among the assemblage of varied views, I found that opinions frequently were not based on facts.

In this book I try to sort out the differences. While I do not conceal my views, I seek to present the positions of all, fully and fairly. My sense of obligation derives from gratitude to the people who shared their thoughts with me and for concern that the wisest policy be adopted by the nation.

I have spoken with about 200 people concerning the radon issue. Some conversations were brief, and others lasted for hours. References in the book to these exchanges are cited as interviews, whether short or long, by telephone or in person. I cannot list by name all the people from whom I benefited, but several were particularly helpful. I am especially grateful to the 10 individuals with varying positions whom I profile in Chapters 3 and 4. I characterize them as leapers, loopers, lopers, and loppers according to their positions on radon policy. Among them, Steven Page, director of the radon division in the Environmental Protection Agency (EPA), and Susan Rose, manager of radon programs in the Department of Energy, were unusually generous with their time and assistance.

I have also gained from discussions with the two former directors of the EPA's radon division, Richard Guimond, now assistant administrator for Superfund management, and Margo Oge, director of the agency's Office of Radiation Programs. Some other federal agency officials, state officials, and congressional aides with whom I spoke are cited by name in the book, though many more are not. In working through the scientific and technical issues, I profited from discussions with several scientists. I thank Naomi Harley, John Harley, and Jan Stolwijk in particular for clarifying issues concerning the science of radon described in Chapter 2.

Richard Benfield, editorial page editor of *The Record* of Hackensack, New Jersey, provided access to his newspaper's data base, which helped immeasurably to trace his paper's treatment of the radon issue. I gained much from people associated with radon programs in other countries, including Ernest Létourneau in Canada, Gun Astri Swedjemark and Lynn Hubbard in Sweden, and Olli Castrén in Finland. I am grateful to Tuomo Martikainen, head of the political science department at the University of Helsinki, where in the summer of 1991 I was a visiting research scholar.

Sheldon Krimsky and Joan Aron made valuable comments about the manuscript in draft form. I am greatly indebted to Joseph Morone for his early encouragement and cogent suggestions as the book developed.

In deference to readers who are unfamiliar with the vocabulary of radiation physics, I have tried to keep simple the complicated mix of scientific terms, measures, and abbreviations. The world of radiation units—curies, becquerels, rads, working levels—can be bewildering to a nonspecialist. In the course of the book I refer to these terms when necessary, but always with explanations in consideration of the nontechnical reader.

Fortunately, terminology associated with radon measurement does not require understanding the array of radiation units. Quantities of radon are commonly described in units called picocuries. The usual terminology is "picocuries of radon per liter of air," abbreviated pCi/L. In the text, I use the abbreviation sparingly. Rather, I spell out the full term, or write "picocuries" (leaving tacit the phrase, "of radon per liter of air") when the meaning is otherwise clear.

This approach reflects a goal of the book: to enhance its literary flow and reach a readership beyond the science community. Radon policy can affect the lives and pocketbooks of every citizen, scientist or not. Nonexperts therefore deserve to know what scientists and policy makers know about radon and what they do not know. It is in that spirit that I offer this exploration of the politics of radon and its implications for the broader subject of science and public policy.

Leonard A. Cole

Ridgewood, New Jersey
January 1993

CHAPTER 1

The Development of Radon Policy

RADIATION, RADIOACTIVITY—WORDS THAT CAN CONJURE fear and fascination—are the raw material of this study. They are the phenomena associated with radon gas and its decay products that have prompted government authorities to voice alarm about the ostensible hazard the gas poses to millions of Americans.

Official concern arose in the 1980s when elevated concentrations of radon were found in homes across the country. At times, reports of government pronouncements conveyed a sense of crisis. "Major Radon Peril Is Declared by U.S." said a *New York Times* front-page headline in 1988.[1] The story reported that in a survey of seven states, the Environmental Protection Agency (EPA) had found many homes with higher levels of the gas than expected. This prompted Lee M. Thomas, the agency's administrator, to urge that every home in the United States and every apartment from the second floor down be inspected for radon. At the same time Dr. Vernon J. Houk, Assistant Surgeon General of the Public Health Service, declared that "radon-induced lung cancer is one of today's most serious public health issues."[2]

Although generally out of public view, scientists had begun to debate the extent of the danger, the manner of addressing it, and the proper role of government in the process. Some experts contended that the government was not acting with sufficient vigor to protect the public. Others argued that the gas posed virtually no risk to the general citizenry and that no indoor radon policy was warranted. This book is about these views and their wisdom. It is based on government publications and hearings, scientific reports, media coverage, and interviews with interested parties, including scientists, government officials, and spokespersons for business and environmental organizations.

The issue is not trivial from either a health or financial perspective. Scientific models suggest that among the estimated 130,000 annual lung cancer deaths in the United States, as many as 50,000 could be attributable to indoor radon exposure.[3] The EPA presumed in 1986 that the annual number of radon-induced lung cancer deaths is between 5,000 and 20,000.[4] At the same time it advised that if radon concentrations exceeded 4 picocuries of radon per liter of air, mitigation should be performed. (A curie, named after the French physicists Marie and Pierre Curie, is the amount of radioactivity emitted by the decay of a gram of radium. A picocurie is one trillionth of a curie.)

The EPA's advisory to test for radon and to reduce the levels to below 4 picocuries would cost Americans between $8 billion and $20 billion.[5] If a federally legislated goal for the country were implemented, the expense would be astronomical. The 1988 Indoor Radon Abatement Act declared as the long-term national goal that all buildings in the United States "should be as free of radon as the ambient air outside of buildings."[6] Several scientists doubted that the goal was technically possible to achieve, but if it were, it would cost $1 trillion.[7]

Democracy and Science Policy

Beyond the content of the radon debate lies the issue of how it should be presented to the American people. If there is common agreement about the hazard of a pollutant or chemical, how far should government go toward reducing the hazard? The question is pertinent, for example, to the issue of smoking, which is universally regarded as a health risk. Should the government ban smoking entirely? Should it forbid tobacco only in enclosed places where the public gathers? Should it merely provide information and let individuals decide for themselves whether and where to smoke? In the context of individual freedom, forbidding smoking in public buildings seems legitimate because of the discomfort and health effects tobacco smoke may have on nearby nonsmokers. But in a society that cherishes personal freedom, it is not inconsistent to allow smoking in private despite personal risks to health.

If the government's proper role about smoking policy remains unclear, indoor radon provides an even greater challenge, because its presumed hazards are less certain. Democratic principles suggest at the least that government is obliged to provide a citizenry with the best information available. Only then can people make rational decisions about their personal behavior (or irrational ones if they care to). In a word, the public is entitled to know the range of views on an issue, presented in a fair and open forum.

In the case of radon policy, tension has evolved about this basic democratic premise. Much of the public has ignored EPA's advisories to test and remediate their homes. By 1992, despite six years of urging by the EPA, only about 5 percent of the nation's homes had been tested.[8] Frustration was evident at a 1991 EPA-sponsored international radon symposium. The keynote speaker, John R. Garrison, managing director of the American Lung Association, cited the "partnership" of the EPA with his organization and other "concerned agencies." Indeed he noted that the EPA had given the lung association money to promote awareness about radon. He suggested, however, that the public still lacked sufficient "respect" for the dangers of radon. In consequence, he said to the assembled scientists, physicians, and educators, "[i]n your communications with the media and with the public, I urge you not to dwell on the doubts over radon, but on the legitimate dangers of long-term exposure." Disparate opinions about radon and its health consequences should be muted for "we need to speak with one voice to the public."[9]

In effect, the conference keynote was an appeal that uncertainties about the wisdom of the EPA's policy be hidden from the public. This book is dedicated to the opposite premise: A healthy democracy requires that uncertainties and disparate views be laid before the public. Not only does this study reveal the assortment of views on radon, it examines the manner in which American institutions—government agencies, the press, the scientific community—have conveyed information about radon to the public. Frequently, as will be discussed, these institutions have been found wanting.

Apart from openness, a second plea is made here for civility. Debates about radon policy have often been mired by innuendo and charges of misrepresentation. Neither supporters nor critics of the government's policy, including EPA officials, have abjured such behavior. When *New York Times* writer Jane Brody reported in January 1991 that "several reputable scientists say that the extent and seriousness of radon contamination have been greatly exaggerated," she quickly heard from the EPA.[10] Margo Oge, director of EPA's radon division, charged that Brody had written about "the outlying opinions of a few scientists." The article, Oge said, was "a rehash of the same old story from the same critics."[11]

One of the scientists cited in Brody's article was Naomi Harley, a research professor at New York University's Department of Environmental Medicine. A recognized authority on radon, Harley received her own letter from Oge in April. The EPA radon director wrote to her about published comments concerning Harley's radon studies. The studies showed that EPA's recommendations to the public may have exceeded prudential needs. Oge wrote: "I do not believe that your remarks as reported in the press late last week do much to give people an improved understanding of radon." She accused Harley of "misreporting" EPA's position and said that press accounts showed Harley "suggesting that radon is not a health concern."[12]

Distressed by Oge's letter, Harley replied that "[b]ecause I do not accept the exaggerated risk estimates of EPA does not indicate that I 'suggest that radon is not a health hazard.' I urge you not to misrepresent my publications and committee work on radon."[13]

The EPA was not taking kindly to its critics. But neither would all its critics mute their scientific judgments and heed the entreaty to "speak with one voice." Indeed, criticism by some bordered on invective. In a communication to the EPA's radon program review panel, Anthony Nero, a senior scientist at Lawrence Berkeley Laboratory in California, offered a scathing critique of the EPA program. A researcher and prolific writer on the subject, he presented what he considered to be a proper radon policy (which will be outlined later in the book). Spliced into his proposals were accusations that the EPA had engaged in "sleight of hand," displayed "notable arrogance," issued "inaccurate and misleading" information, and "misled and manipulated the public."[14]

To be sure, the EPA was condemned as well by people who thought its policies were insufficiently aggressive. Gloria Rains, an environmental activist, criticized the EPA's action level of 4 picocuries of radon per liter of air as allowing for "an excess fatal lung cancer risk of 3-in-100 [which] is unconscionable."[15] While she was urging that the action level be lowered, Philip Abelson, a distinguished scientist and editor of the journal *Science*, thought the entire radon program should be eliminated. The EPA, he said, was trying to "brainwash" the public into believing there was a radon problem where none exists.[16]

Thus, the radon issue has suffered from vituperation by those who make or wish to influence policy. Intemperate language can only introduce emotion that inhibits rational discourse. The consequent tensions, disquieting in their own right, are all the more upsetting as they warp discussion about an issue that could be affecting people's lives.

The EPA position on radon has in fact been generously provided to the public. Despite occasional news articles like Jane Brody's, the public has infrequently been exposed to contrary views in the media. Similarly, as will be shown in later discussion, supporters of the EPA's position are more likely to be called to testify at congressional hearings than are skeptics. As a result, scientists have expressed dismay that the debate before the public and its elected representatives has been imbalanced. The following pages will expound on these contentions and attempt to provide all sides of the issue, including uncertainties about the assumptions behind current radon policies.

Apropos of this theme is a recent news story about EPA's policy concerning dioxin. In 1982 the federal government bought out the town of Times Beach, Missouri, and permanently evacuated its 2,242 inhabitants. The action was taken after the discovery that dioxin had been spread on the town's roads to control dust. In 1991 a federal health official who played a central role in the 1982 decision acknowledged that the danger from the chemical had been exaggerated, and he regretted his part in urging the town's evacuation: "We should have been more upfront with Times Beach people and told them, 'We're doing our best with the estimates of risk, but we may be wrong.' I think we never added, 'But we may be wrong.'"[17]

The federal official was Vernon Houk who, as noted earlier, in 1988 announced unequivocally that "radon-induced lung cancer is one of today's most serious public health issues" and in 1990 said to a Senate subcommittee that his "testimony can be summarized in one sentence. Indoor radon poses a very significant health threat, and the country should get on with doing something about it."[18] No room for uncertainty there.

The risk usually associated with radon relates to its effect on health. But there is another element of risk—that citizens will become disillusioned with authorities whose policies prove to have been unwarranted. The premise in this book is that Houk's belated wisdom about government policy on dioxin in 1982 should inform radon policy in 1993: "But we may be wrong." To that end, this book lays out the politics and the science behind current radon policies. The guiding proposition here is that more, not less, information will lead to the best possible understanding of the issues.

Radon Policy and Human Values

Thousands of articles and papers have been written about radon, most of them scientific reports. They include studies about radon's physical and chemical characteristics, its distribution patterns, and its biological effects on humans and animals.[19] In recent years several full volumes about radon have appeared. Most are technical studies aimed at scientists[20] or consumer-oriented books that urge people to test their homes for the gas and remediate as necessary.[21]

An increasing number of studies on risk analysis have dealt with radon. They range from attempts to assess risk exposure to humans by mathematical models to studies that tie risk communication techniques to public action.[22] Although some writings on radon make reference to public policy, few have inquired into the political tensions that lie beneath the surface of current radon policy. The only lengthy exposition on the social implications of the policy appears in a chapter of Sheldon Krimsky and Alonzo Plough's *Environmental Hazards: Communicating Risks as a Social Process*.[23] But even here, the authors demur from judging the merits of U.S. radon policy or the manner of its development. Rather, they discuss radon and other risk issues in the context of risk communication.

Nevertheless, Krimsky and Plough's emphasis on the importance of "cultural rationality" lends perspective to this study. Their exploration of culture and human values in understanding people's responses (or lack of responses) to a technically defined risk reminds us that the experts who make technical assessments are also people. As do nonexperts, they bring emotion and biases to their activities. That is often the reason that reputable scientists have profound disagreements among themselves about supposedly objective scientific evidence.[24] The value-laden dimension is central to understanding the policy process.

Science policy formulation begins with concrete facts. The facts, however, are subject to judgments and interpretations, which become the scaffolding of public policy. Thus it is an indisputable fact that human exposure to certain levels of radiation increases the risk of cancer. But determining precisely what the levels are and whether some humans are more susceptible than others remains in dispute. Out of the mix of facts and judgments comes public policy.

Despite ample publicity during recent years about radon's supposedly widespread threat, the public seems remarkably unworried. This confounds a traditional assumption that people are more concerned, sometimes irrationally, about radiation than other environmental hazards.[25] The apathy is frequently attributed to radon's being a natural rather than man-made source of radiation. Henry Wagner and Linda Ketchum, for example, write that radon does not "produce the same feelings of alienation from authority, vulnerability to the capricious actions of others, and impotence that nuclear energy seems to produce. The problem is 'natural,' so there is no one to blame."[26]

The explanation is incomplete, as will be discussed in the book, but it underscores the importance of values in making policy decisions. It is a reminder that circumspection is preferable to certitude, that scientists who advocate contradictory positions cannot all be arguing "objective" science. Indeed, in the case of radon, much of the basic science is not in dispute, yet interpretations and policy recommendations vary widely.

The radon issue affords an unusual opportunity to examine the flowering of a policy ostensibly drawn from scientific knowledge. Its recency allows for discussion with scientists, government authorities, and others who have been involved since the policy's inception. Rich not only in scientific overtones, the issue raises profound questions about the value this society attaches to health and environmental protection.

Radiation

The scientific core of the radon issue is radiation and its effects on health. Of particular concern is radiation that causes ionization, a process by which atoms become electrically charged. Ionizing radiation, or radioactivity, releases high levels of energy. The release of energy occurs when the radiation waves or particles collide with a stable atom, thus causing the loss of one or more electrons from the atom. An electrically charged atom, called an ion, is likely to interact with other particles or atoms.

Under controlled conditions, the process enhances human health and welfare. Medical x-rays are nearly indispensable diagnostic tools, and research in many areas of biology, chemistry, and physics is dependent on the use of radioactive elements. As beneficial as radiation may be under certain conditions, in others it is harmful. Radiation may destroy living cells or alter their structures so they become cancerous. Even this capability has been put to human use, for example, in the therapeutic application of high levels of radiation to destroy cancer cells. But uncontrolled exposure to high-level radiation is universally understood to be life threatening.

A phenomenon as old as the universe, radiation has been an ever-present part of Earth. Atoms disintegrate continuously both at the core of the planet and above the surface—everywhere, all the time. All manner of being, inorganic and organic, has evolved in a bath of radiation. Indeed, without radiation there could have been no life, just as with too much radiation there can be no life.

The science of radiation began nearly a century ago with three

remarkable discoveries. In 1895, Wilhelm Roentgen found that passing an electric current through a glass bulb from which most of the air
had been removed caused a strange reaction. It generated rays that
created penetrating images on certain chemically sensitive material,
such as a photographic plate. When directed at a hand, for example,
the rays would pass through the soft outer tissue but not the denser
bones, whose images could be seen on the plate. He named these
mysterious rays x-rays.[27]

The following year, Henri Becquerel discovered that rays from
uranium could produce similar penetrating images. Marie Curie pursued work on spontaneously emitted radiation, which she named
radioactivity. In 1898, she and her husband Pierre identified an ingredient of uranium ore that was unusually active in the emission of rays.
They had discovered radium.[28]

Radioactivity was greeted by the world with fascination and little
sense of danger. X-rays were quickly put to use by physicians, and
radium was seen as an "elixir of life," with curative powers for a variety
of ailments.[29] In subsequent years, the potential for radiation to cause
damage became increasingly apparent. Consequently, radioactivity
assumed an aura that continues to the present, a confusion of promise
and fear. For some, as Spencer Weart recounts, the release of energy
within atoms was "a guarantee of all the wonders that modern civiliza-
tion would inevitably bring."[30] But radioactivity also carried a pall of
anxiety with the knowledge that "a power great enough to transfigure
flesh could destroy it."[31]

Radon

Radon gas is a decay product of radium. It lies in the radioactive decay
chain that begins with uranium-238, an element whose half-life (the
period during which half the material decomposes) is 4.5 billion years.
Although uranium is particularly abundant in granite, shale, and phos-
phate-bearing formations, small amounts are dispersed throughout
the earth's crust. Radon therefore may be found everywhere.

Since it is a gas, radon filters through cracks in the bedrock and
soil before it finally escapes into the atmosphere. Its half-life is only
3.8 days, and its decay products, collectively called radon daughters or
progeny, are themselves radioactive elements with half-lives ranging
from a fraction of a second to 22 years. (The radon decay chain's
elements, their half-lives, and the nature of their emitted radiation are
shown in Appendix A.)

When inhaled, radon gas flows quickly in and out of the lungs, almost never lingering long enough to cause damage. But the radon progeny, which are solids, tend to lodge in the bronchial tree. Here they emit "heavy" alpha particles, smaller and lighter beta particles, and short-wavelength gamma rays. Of the three kinds of emission, alpha particles are the most harmful; because of their greater electrical charge and relatively large mass, they can cause more damage to tissue. Skin is an effective barrier against them, because they can pass through only one or two cells before being brought to a halt. But the lungs lack such a protective layer, and so the particles can affect sensitive basal cells lining the bronchi when radon progeny become lodged there. Alpha particles can damage the DNA in cell nuclei, and that damage can lead to uncontrollable cell reproduction and the growth of a cancerous tumor.

Thus, there is no question that radon can pose health hazards. Its harmful effects on people have been observed at least since the mid-sixteenth century—almost 350 years before it was identified as the culprit. In 1556 the physician Georgius Agricola found that lung disease was so prevalent among silver miners in the Erz Mountains (on the border between Germany and Czechoslovakia) that some women in the area lost as many as seven husbands to premature death. F.H. Harting and W. Hesse first described the malady as cancer in 1879, after clinical and autopsy examinations revealed intrathoracic neoplasms in many miners. Subsequent reports indicated that between 1875 and 1912, as many as half the men working in the mines eventually died from lung cancer.[32]

By the 1930s some physicians suspected a relation between lung cancer and radioactivity in the mines. The ore contained high concentrations of uranium, and the air contained high concentrations of radon. Other observers, however, rejected the radiation hypothesis and blamed miners' inbreeding for a genetic predisposition to the illness. In the 1940s, some scientists still argued that radon alone was not the cause of lung cancer in miners, but that their genetic susceptibility was high.[33]

With the dawn of the nuclear age uranium mining increased, and along with it, the number of miners succumbing to lung cancer. At the same time, the biological basis of carcinogenesis became better understood. Ultimately, epidemiological studies of miners in Europe and the western United States during the 1950s and 1960s confirmed that prolonged exposure to high concentrations of radon gas in mines does cause lung cancer.

In the late 1960s, after the link between radon and cancer had been acknowledged, if not fully understood, the Public Health Service set a limit on the amount of the gas to which any American miner could be exposed: 100 picocuries per liter of air. This regulated occupational limit is commonly referred to as a working level. In fact, a working level is a measure of radioactivity of radon's decay products rather than of the gas itself. A technical definition of a working level is "Any combination of short-lived radon daughters [decay products] in one liter of air that will result in the emission of 1.3 X 10^5 MeV [megavolts] of potential alpha energy."[34]

Thus, 100 picocuries of radon in a liter of air will decay into progeny that will in turn decay and, in the process, yield alpha radiation corresponding to one working level. It is the alpha particles deposited in the lungs that can cause cancer. In an unventilated mine, 100 picocuries of radon are considered equivalent to one working level because the gas and its decay products are confined and are deemed to be in "equilibrium." Above the earth's surface, however, the gas dissipates more quickly and its progeny in a given location will be proportionately lower. In most homes and other buildings, the ratio of radon progeny to the gas is considered to be 50 percent of equilibrium. In indoor structures, then, 100 picocuries of radon per liter of air would be equivalent to half a working level. The EPA's action level of 4 picocuries of radon gas is equivalent to 0.02 working level of radon progeny.[35]

Emergence of Radon Policy

With the increase in uranium mining since the 1940s came a growth in mills to extract the element from the ore. In the refining process, the mills produced large quantities of a sand-like byproduct called tailings. In the 1950s and early 1960s the radiation-emitting tailings were used as fill material in the construction of houses, schools, and commercial buildings in nearby locations. Government health officials ordered the practice stopped in 1966 after finding that many of the structures had elevated radon levels. The first remedial action program in the United States for indoor radon was undertaken in the 1970s for about 300 of these homes in Grand Junction, Colorado. The plan involved excavation and removal of radioactive fill and replacement by nonradioactive material. In 1978, the average cost per home for remediation was reported to be about $13,500.[36]

Before cleanup, 71 percent of the houses had levels above 6 picocuries of radon per liter of air. After cleanup, although the highest concentrations were reduced, 68 percent still had levels exceeding 6 picocuries.[37] Thus, after removing and carting away the radioactive fill material, more than two-thirds of the "remediated" homes had radon levels above 6 picocuries.

Concern about the health effects on people living in homes above the mill tailings prompted passage in 1978 of the Uranium Mill Tailings Radiation Control Act. The act required the EPA to establish radiation standards for homes involved with the tailings. The agency ultimately complied in 1983. Calling its new standard an "optimized cost-benefit" alternative, the EPA set 4 picocuries per liter of air as an objective, with a maximum limit of 6 picocuries.[38] Ironically, the agency had become aware that houses not on tailings also had elevated radon levels, yet applied no policy to them.[39]

In setting the 1983 limits, the agency frankly acknowledged that it had made assumptions for which there was no demonstrated evidence.

> For the purpose of establishing standards for the protection of health, we assume a linear, nonthreshold dose-effect relationship as a reasonable basis for estimating risks to the general public from radiation. This means we assume that any radiation dose poses some risk and that the risk of low doses is directly proportional to the risk that has been demonstrated at higher doses. We recognize that the data available [do not preclude] a threshold for some types of damage below which there are no harmful effects....[40]

The regulation indicated that the federal government would pay for cleanup, which would be conducted by the Department of Energy (DOE). The relevance of the EPA's 1983 regulation to its future radon policies carries a trace of paradox. The 1983 ruling established a standard. Thus, the goal of 4 picocuries for homes on tailings mandated an action response. It was not just an advisory, as it would be in 1986 for other homes throughout the country when EPA first announced a policy about them. Yet in Grand Junction, most "remediated" homes continued to have radon levels above 6 picocuries. Thus, the rationale of the newly proclaimed EPA target of 4 picocuries seemed questionable.

EPA's policy assessment of the risks has derived in part from models drawn by outside experts. As discussed in the next chapter, while many scientists presume the risk models to be valid, all acknowledge uncertainty. The most striking feature of the 1983 regulation can be seen in retrospect: In no subsequent EPA publication has the agency so forthrightly acknowledged uncertainty about the basis of its policy assumptions. The agency's widely distributed brochure, its 1986 *Citizen's Guide to Radon*, mentions that "there is some uncertainty about the amount of health risk," but offers no elaboration. Rather, in bold print the text emphasizes that "the greater your exposure to radon, the greater your risk of developing lung cancer."[41]

The Watras Incident

Before 1985, U.S. residential radon policy was focused almost exclusively on man-made sources, such as houses built over uranium or phosphate tailings. But a bizarre incident at the end of 1984 catapulted the indoor radon issue to a new plane of concern. Stanley J. Watras was working as a construction engineer at the nearly completed Limerick nuclear power plant in Pottstown, Pennsylvania. Early in December 1984, weeks before the plant was to become active, radiation monitors were installed through which workers had to pass. Since the day the monitors were activated, Watras recalls, "I constantly tripped every alarm on all zones." The Philadelphia Electric Company, which owned and operated the plant, determined that Watras' clothing and skin were contaminated with isotopes of natural background radiation, unrelated to the proposed plant operations.[42]

Increasingly apprehensive, on December 14 Watras traced "every step I made from my home to the job site." He could find no explanation and asked the company to test his house in Boyertown. On December 19, after measuring high radon levels in the living room, the company contacted the Pennsylvania Department of Energy Resources. On December 24, the department installed testing devices throughout the house and left them for a week. On January 5, three days after department authorities removed the devices for analysis, a state official presented a letter in person to the Watras family urging them to move out of their house. Measurements were as high as 2,700 picocuries. The following day the family moved to a motel.[43]

In July, state authorities assured Watras that his home had been "restored to an inhabitable condition," and he and his family returned. Meanwhile, tests in some 2,900 homes in the surrounding area indicated that about 40 percent had "above acceptable" radon levels.[44] Though none was nearly as high as in the Watras house, authorities became convinced that elevated indoor radon was more extensive than previously supposed. EPA and DOE officials announced plans to assess radon concentrations and possible health consequences around the country. (As will be discussed, each agency had begun research programs on indoor radon in the 1970s.) Investigations would focus around geological formations such as the Reading Prong, an area rich in uranium ore that extended through parts of Pennsylvania, New Jersey, and New York.[45]

Role of the Environmental Protection Agency

"The Watras incident really changed the ballgame," recalls Richard Guimond, who became the first director of EPA's newly established

Radon Division in 1986.[46] Preliminary findings in 1986 led the EPA to conclude that perhaps 10 percent of the nation's 80 million homes had radon levels above 4 picocuries per liter of air.[47] Moreover, many of these homes were not in predictably high-uranium areas. The EPA's estimate was reported under somber headlines such as *Newsweek's* "Radon Gas: A Deadly Threat—A Natural Hazard Is Seeping into 8 Million Homes." The article cited politicians and environmentalists who criticized the EPA for being "all talk and no action" and "not telling the public how high the risks are."[48]

In August the agency issued two brochures. The first brochure, *Radon Reduction Methods: A Homeowners Guide*, was a 23-page guide that explained ventilation and sealing techniques. The booklet readily admitted that knowledge about mitigation methods was "far from complete." The information was for "those whose radon problems demand immediate attention."[49]

The second brochure, *A Citizen's Guide to Radon: What It Is and What to Do About It*, was less circumspect. More than a million copies were issued, making it by far the most broadly disseminated government publication on the subject. Apart from glossing over uncertainties about projections of lung cancer deaths from indoor radon, another crucial feature was out of balance: The brochure virtually disregarded the combined effects of smoking and radon.

On the first page, and without qualification, the booklet said that "scientists estimate that from 5,000 to about 20,000 lung cancer deaths a year in the United States may be attributed to radon." It then noted that around 85 percent of the 130,000 expected lung cancer deaths in the United States in 1986 were attributable to smoking.[50] But the EPA never explained the gap in logic suggested by these observations. According to the figures, 15 percent, or around 20,000 people, die of lung cancer from nonsmoking causes. But if the EPA's upper limit of 20,000 radon-connected lung cancers is correct, no room is left for the many other causes of lung cancer, like passive smoking, asbestos, aromatic hydrocarbons, and other chemicals and pollutants.

The booklet included a phrase on the next to last page that "smoking may increase the risk of exposure to radon."[51] But the words are tentative and so distant from the lead message that they seem an unimportant conjecture. In fact, an apparent synergism of radon and smoking means that most ostensibly radon-caused deaths could be eliminated if smoking were eliminated.

The increased concern about indoor radon also prompted legislative action. In October 1986, Congress enacted a Superfund amendment called the Radon Gas and Indoor Air Quality Research Act.[52] The act in effect designated the EPA administrator chief of radon policy in the United States. That person would be in charge of "research, development, and related reporting, information dissemination, and coordination activities" concerning radon policy. In essence the EPA was assigned the task of finding out the extent of hazard posed by indoor radon and how to address the problem. Nowhere in the bill is the DOE mentioned, although the department had been engaged in radon research for years.

In 1988, Congress enacted a far more comprehensive radon bill, the Indoor Radon Abatement Act. Like the 1986 act, this statute also ignored the DOE's role in radon activities, and solidified the role of the EPA as the principal locus of radon policy formulation. The EPA administrator was instructed to publish revised versions of the *Citizen's Guide* with updated information about health risks, feasibility of reducing indoor concentrations, and the relationship between short-term and long-term testing techniques. In addition, the administrator was to develop construction techniques to control radon in new buildings, assist states with radon programs, assess the extent of radon in the nation's schools, and designate regional radon training centers in universities around the country. For these activities the act allocated more than $30 million beyond the millions already budgeted for EPA's radon work and its 90-member radon staff.

The act's most controversial provision was its opening statement.

> The national long-term goal of the United States with respect to radon levels in buildings is that the air within buildings in the United States should be as free of radon as the ambient air outside of buildings.[53]

Despite doubts about whether the goal was technically possible to accomplish, supporters insisted that the objective was appropriate to strive for.

Meanwhile, the EPA had become concerned about the public's lack of response to its radon advisories, although media coverage of the subject was considerable. As a result, the agency's public information efforts became more aggressive and, in the process, minimized the scientific uncertainties that EPA had freely cited in earlier pronouncements.

In 1989, the EPA sponsored a publication called *Reporting on Radon*, subtitled *A Journalist's Guide to Covering the Nation's Second-Leading Cause of Lung Cancer*. The publication was issued by the National Safety Council "in cooperative agreement" with the EPA. The 64-page booklet was replete with hyperbolic imagery. The EPA's previous 5,000-to-20,000 estimate of annual lung cancer deaths from radon gave way only to the higher figure. "Some 20,000 Americans die each year because of radon-induced lung cancers, the Environmental Protection Agency now estimates."[54] Radon decay products are "deadly threats" and "tiny time-bombs" that "go off" in the lungs. Alpha particles "slam into unshielded lung cells."[55] The booklet's information was sometimes plainly wrong, as in its unqualified assertion that "children are more sensitive to radon exposures than adults."[56] Scientific studies place this assumption in doubt.[57]

Reporting on Radon explains the radon issue in simple language and does cite scientists who think the EPA's risk estimates are exaggerated. It differentiates between "responsible" critics and those who hold the "'extreme' radiation-is-good-for-you position."[58] But it says nothing more about the "extreme" view, and the reader is left wondering who the irresponsible critics are. In a larger perspective, the booklet represented another step away from hesitancy by the EPA about its assumptions. A 1990 EPA-sponsored campaign in conjunction with the Advertising Council went further and abandoned all doubts.

The EPA/Ad Council Campaign

Nothing the EPA has done regarding radon has generated more criticism than the campaign it developed with the Ad Council. The council describes itself as "a private, non-profit organization of volunteers who conduct advertising campaigns in the public good." When the EPA approached the Ad Council for help in 1988, the council agreed because "we believe radon is a significant health problem, a problem that can be addressed in part via strong informative public service advertising."[59]

The campaign included distribution of materials such as pamphlets, balloons, bumper stickers, caps, and t-shirts emblazoned with messages about the danger of residential radon. Since 1989, EPA/Ad Council advertisements have appeared on billboards, television, and in the print media. Jeffrey Boal, an Ad Council supervisor who helped develop the radon campaign, said that through 1991 the EPA had spent about $2.5 million on production of materials.[60] By 1992, media time worth another $75 million had been provided pro bono.[61]

Boal characterized the campaign as aggressive although, he acknowledged, others considered it alarmist. The campaign material revealed that the EPA had shed any hint of uncertainty about the validity of its assumptions and action levels. A newspaper advertisement urged readers to call a hotline, 1-800-SOS-RADON, for information, "Because what you don't know can do more than hurt you. It can kill you." A television sequence showed a family at home being transformed into skeletons, evidently caused by exposure to radon. (Appendix C includes pictures from the television advertising.) A pamphlet cover asked in bold red print: "Has your home been invaded by Radon?" The question is followed by an ostensible reassurance: "If so, don't panic. Act now." Presumably no one was panicky before reading the question.

The pamphlet's inner fold includes a risk evaluation chart indicating that if 100 people were exposed to an annual radon level of 4 picocuries per liter of air, about two people among them "may die from Radon." Nothing is said about the number of years the people would have to have been exposed. In fact, EPA's earlier conjectures were all premised on individuals spending 75 percent of their time in a home for 70 years.[62]

Going beyond the 1986 *Citizen's Guide* advice to remediate at 4 picocuries of radon per liter of air (4 pCi/L), the pamphlet implicitly lowered the action level. A boxed notice says: "In most cases, you can reduce the Radon level in your home to as low as 2 to 4 pCi/L, and sometimes even below 2 pCi/L." The EPA seemed to be tracking in the direction of the national goal to make indoor levels as low as those outdoors (about 0.5 picocuries per liter of air).

Another EPA/Ad Council pamphlet warns: "Protect your family against Radon—the silent killer." It contains a picture of a chest x-ray with the caption: "Having Radon in your home is like exposing your family to hundreds of chest x-rays yearly." Nothing in the caption suggests that some radon is unavoidably in everybody's home. The pamphlet advises that "your home be tested *immediately*" (original emphasis). In contradictory signals it says: "Your family's risk of developing lung cancer from Radon depends on the average annual level of Radon in your home." While saying in italics that long-term testing is the most accurate way to test for radon, it also emphasizes that *"[s]hort-term testing (a few days to several months) is the quickest way to determine if a potential problem exists"* (original emphasis). As will be shown in the following chapter, many scientists think short-term testing can be misleading.

Despite criticism of the EPA/Ad Council approach by many in the scientific community, an EPA spokesperson described it as "an innovative and outstanding campaign which has greatly enhanced the public's recognition of radon as a problem."[63]

The 1992 Citizen's Guide to Radon

In May 1992 the EPA issued a second edition of *A Citizen's Guide to Radon*. While not explicitly a product of the Advertising Council, the revised brochure reflected the approach developed in the campaign. With a new subtitle—*The Guide to Protecting Yourself and Your Family from Radon*—it is replete with cartoon-like drawings of happy families, including pets, in and around their homes. Included also is a drawing of a home whose rooms are being filled with radon gas, depicted as wavy red lines seeping in from below. Another shows an outline of a person evidently breathing in radon—portrayed as red dots entering the nose and mouth, passing through the bronchi, and spreading out of control in the lungs.[64]

The lead page of the 15-page brochure peremptorily says: "Fix your home if your radon level is 4 picocuries per liter (pCi/L) or higher." Departing from the 1986 *Citizen's Guide*, the new version then suggests that homeowners seek even lower levels. "Radon levels less than 4 pCi/L still pose a risk, and in many cases may be reduced."

Unlike the earlier *Guide*, the 1992 version acknowledges a risk relationship between smoking and radon near the beginning of its text. On page 3 it says in bold print: "If you smoke and your home has high radon levels, your risk of lung cancer is especially high." The statement receives no elaboration until page 12. Here charts suggest that at various levels of radon above 2 picocuries per liter, the risk of contracting lung cancer is about 15 times greater for a smoker than for a person who never smoked.

The only reference to uncertainty appears on page 11, with the comment that "[l]ike other environmental pollutants, there is some uncertainty about the magnitude of radon health risks." But the question is minimized by informing the reader that information from miners' experiences tells us "we know more about radon risks than risks from most other cancer-causing substances." The 1992 *Citizen's Guide* concludes as it begins, with admonitions shorn of doubt. "You will reduce your risk of lung cancer when you reduce radon levels," and "radon levels below 4 pCi/L still pose some risk."[65]

Role of the Department of Energy

As the EPA grew into the nation's chief policy-making body on radon, it became more involved in applied research as well. The Department of Energy, however, remained the principal sponsor of basic research on the gas. With an annual budget for radon projects of about $13 million in 1990, the following year the DOE sponsored 68 investigations. In the 3-year period from 1988 through 1990, more than 500 scientific publications resulted from DOE-supported studies.[66] The projects ranged from research on radon seepage into buildings to lung cancer risk and the mechanisms of radon carcinogenesis.[67]

The DOE's radon program is conducted under the auspices of its Office of Health and Environmental Research. The work is an extension of decades of research that the office and its predecessor bodies had conducted on the biological effects of ionizing radiation. The lines of current work can be traced to the first U.S. studies of radon in mines in 1951 and in home environments in 1976.[68]

The DOE and EPA differ not only in research areas, but in policy emphases. While the EPA has largely shaped national policy on indoor radon, DOE publications are less aggressive. They stress uncertainties about current scientific knowledge and implicitly call for a more restrained national policy. The DOE's 1991 radon research program study, for example, says: "There still remains a significant number of uncertainties in the currently available knowledge that is used to estimate lung cancer from exposure to environmental levels of radon and its progeny."[69] The contrast with EPA's warnings to protect against the "household intruder" and to test "immediately" was striking.

A 1989 DOE report on radon and epidemiology begins with the observation that "the issue of public health risk from indoor radon has grown in both importance and controversy." It continues: "Significant scientific questions yet remain about extrapolating the lung cancer risk from radon exposure in uranium miners, the source for radon estimates, to that of the general public."[70] The report covered the proceedings of a DOE-sponsored workshop on all known projects throughout the world involving residential radon epidemiology. Although most of the studies remained incomplete, 18 participants from 10 countries tried to assess the current level of knowledge. They divided into working groups on different aspects of radon research. One was a policy group that identified questions intended to be of interest to policy makers. The questions included the following:

- How serious is the lung cancer risk associated with indoor radon? In homes? In schools? In workplaces?
- What are the risks at different levels of exposure?
- How is the risk modified by smoking? By other factors?
- Are children at greater risk?
- How can radon risks be communicated with perspective?
- Is it a government's role to overcome public apathy?[71]

Workshop participants expressed hope that studies under way might help with answers. Their tone was interrogative rather than declarative. And that implicitly delineates the differing philosophical approaches of the DOE and EPA on the subject of radon. An anonymous federal bureaucrat described the differences as "the DOE having sense but no heart; the EPA heart but no sense."[72]

The Energy Department is reluctant to engage the entire American population in a costly radon program pending clearer evidence that low-level exposure poses a threat to millions. The EPA assumes the worst possibilities (which *are* real possibilities) and, despite incomplete evidence, presses for a nationwide popular response.

These disparate views are not yet reconciled, which returns us to the question raised at the beginning of the chapter: whether uncertainties should be spread before the public, or whether only the worst-case possibilities should be. Neither party's good intentions are in question. Indeed, a beneficence aimed at protecting people from harm, which is the basis of the EPA's approach, deserves recognition as noble in purpose. But good intentions do not necessarily translate into wise policy.

In essence, the issue is about paternalism versus individual freedom—the proper role of government in relation to the citizenry. It is central to a study about the politics of radon.

Notes

1   Philip Shabecoff, "A Major Radon Peril Is Declared by U.S. in Call for Tests," *New York Times*, 13 Sept. 1988, A-1.

2   Ibid.

3   David Bodansky, "Overview of the Indoor Radon Problem," in *Indoor Radon and Its Hazards*, eds., David Bodansky, Maurice A. Robkin, and David R. Stadler (Seattle: University of Washington Press, 1987), 12.

4   U.S. Environmental Protection Agency, *A Citizen's Guide to Radon: What It Is and What to Do About It* (Washington, DC: U.S. Environmental Protection Agency, Aug. 1986), 1. By the early 1990s, the agency had raised the estimate to between 7,000 and 30,000 annual lung-cancer deaths as a result of exposure to radon, according to Michael H. Shapiro, EPA's deputy assistant administrator for air and radiation. See House Subcomm. on Transportation and Hazardous Materials of the Comm. on Energy and Commerce, *Hearing on Radon Awareness and Disclosure*, June 3, 1992 (Washington, DC: Government Printing Office, 1992), 30.

5   Richard Guimond, when director of radiation safety programs in the Environmental Protection Agency, estimated the cost at $8 billion. Anthony Nero, a radon expert at the Lawrence Berkeley Laboratory, said the amount could be as high as $20 billion. They offered these figures at a science writers' workshop on "Radon Today: The Science and the Politics," sponsored by the U.S. Department of Energy in Bethesda, MD, Apr. 25–26, 1991.

6   Indoor Radon Abatement, Title III, Amendment to the Toxic Substances Control Act, Pub. L. No. 100-551, §744, Oct. 28, 1988.

7   A.V. Nero, A.J. Gadgil, W.W. Nazaroff, and K.L. Revzan, *Indoor Radon and Decay Products: Concentrations, Causes, and Control Strategies* (Washington, DC: U.S. Dept. of Energy, Nov. 1990), 108. Philip H. Abelson estimated that

the average cost to homeowners would be about $10,000. "Uncertainties about Health Effects of Radon," *Science*, Vol. 250, No. 4979 (Oct. 19, 1990), 353.

8   Senate Subcomm. on Superfund, Ocean, and Water Protection of the Comm. on Environment and Public Works, *Hearing on Pending Radon and Indoor Air Legislation*, May 8, 1991 (Washington, DC: Government Printing Office, 1991), 4 (statement by Michael Shapiro, Environmental Protection Agency). A bill enacted by the House of Representatives in 1992 states that fewer than 4% of homes had been tested for radon. H.R. 3258, Radon Awareness and Disclosure Act of 1992, *Congressional Rec.* H9698 (Sept. 29, 1992).

9   John R. Garrison, Managing Director, American Lung Association, Keynote Address, Environmental Protection Agency, *The 1991 International Radon Symposium on Radon and Radon Reduction Technology*, Philadelphia, PA, Apr. 2, 1991.

10   Jane E. Brody, "Some Scientists Say Concern Over Radon Is Overblown by E.P.A.," *New York Times*, 8 Jan. 1991, C-4.

11   Letter from Margo T. Oge, director of the radon division, Environmental Protection Agency, to Jane E. Brody, Feb. 27, 1991 (full text in Appendix B).

12   Letter from Margo T. Oge to Naomi H. Harley, Apr. 4, 1991 (full text in Appendix B).

13   Letter from Naomi H. Harley to Margo T. Oge, Apr. 10, 1991 (full text in Appendix B).

14   Letter to Radon Program Review Panel, Environmental Protection Agency, from Anthony V. Nero, Jr., Dec. 10, 1991.

15   House Subcomm. on Health and the Environment of the Comm. on Energy and Commerce, *Hearing on Radon Exposure: Human Health Threat*, Nov. 5, 1987 (Washington, DC: Government Printing Office, 1988), 83 (statement by Gloria C. Rains).

16   Philip H. Abelson, "Radon Today: The Role of Flim-Flam in Public Policy," Science Writers Workshop, 6.

17   "U.S. Health Aide Says He Erred on Times Beach," *New York Times*, 26 May 1991, 20.

18   Senate Subcomm. on Superfund, Ocean, and Water Protection of the Comm. on Environment and Public Works, *Hearing on Radon Testing for Safe Schools Act*, May 23, 1990 (Washington, DC: Government Printing Office, 1990), 5 (statement by Vernon N. Houk).

19   More than 800 studies are referenced in Committee on the Biological Effects of Ionizing Radiation, National Research Council, *Health Risks of Radon and Other Internally Deposited Alpha-Emitters, BEIR IV* (Washington DC: National Academy Press, 1988), and about 1,100 are cited in William W. Nazaroff and Anthony V. Nero, Jr., eds., *Radon and Its Decay Products in Indoor Air* (New York: John Wiley and Sons, 1988). John Harley, a scientist who has worked on radon for many years, has compiled a bibliography of some 2,500 publications on the subject (personal communication, March 11, 1992).

20   Beside the technical studies cited-above, others include NCRP Report No. 78, *Evaluation of Occupational and Environmental Exposures to Radon and Radon Daughters in the United States*, (Bethesda, MD: National Council on Radiation Protection and Measurements, 1984); Philip K. Hopke, ed., *Radon and Its Decay Products* (Washington, DC: American Chemical Society, 1987); NCRP Report No. 97, *Measurement of Radon and Radon Daughters in Air*, (Bethesda, MD: National Council on Radiation Protection and Measurements, 1988); National Research Council, *Comparative Dosimetry in Mines and Homes* (Washington, DC: National Academy Press, 1991). Published symposia and other materials by

government agencies, especially the Environmental Protection Agency and the Department of Energy, will be discussed throughout the book.

21 Bernard Cohen, *Radon: A Homeowner's Guide to Detection and Control* (Mount Vernon, NY: Consumers Union, 1987); David J. Brenner, *Radon Risk and Remedy* (New York: W.H. Freeman and Co., 1989); Douglas G. Brookins, *The Indoor Radon Problem* (New York: Columbia University Press, 1990).

22 Scientific models that lead to differing risk projections about radon are in the studies cited in note [20]. On risk communication, see National Research Council, *Improving Risk Communication* (Washington, DC: National Academy Press, 1989). At the 1991 annual meeting of the Society for Risk Analysis, several of the 31 papers on risk communication referred to radon, particularly in a session titled "How Communication about Radon and Ozone Affects Perceptions of Individual Risk."

23 Sheldon Krimsky and Alonzo Plough, *Environmental Hazards: Communicating Risks as a Social Process* (Dover, MA: Auburn House Publishing Co., 1988).

24 Leonard A. Cole, *Politics and the Restraint of Science* (Totowa, NJ: Rowman and Allanheld Publishers, 1983).

25 Kai Erikson, "Toxic Reckoning: Business Faces a New Kind of Fear," *Harvard Business Review* (Jan./Feb. 1990), 118–26.

26 Henry N. Wagner, Jr., and Linda E. Ketchum, *Living with Radiation* (Baltimore: Johns Hopkins University Press, 1989), 143.

27 Catherine Caufield, *Multiple Exposures, Chronicles of the Radiation Age* (Chicago: University of Chicago Press, 1989), 3–4.

28 Ibid., 22–23.

29 Spencer R. Weart, *Nuclear Fear, A History of Images* (Cambridge, MA: Harvard University Press, 1988), 36–43.

30 Ibid., 12–13.

31 Ibid., 43.

32 Committee on the Biological Effects of Ionizing Radiations, *BEIR IV*, 445; Leonard A. Cole, "Much Ado about Radon," *The Sciences*, Vol. 30, No. 1 (Jan./Feb. 1990), 20.

33 Committee on the Biological Effects of Ionizing Radiations, *BEIR IV*, 446.

34 NCRP Report No. 78, 167.

35 The occupational exposure limit for radon is presumed to correspond to about 20 picocuries per liter of air in homes (averaged for a year). This takes into account the equilibrium factor and differences in time spent at work and home.

36 National Technical Information Service, *Indoor Air Quality Environmental Information Handbook: Radon* (Washington, DC: U.S. Dept. of Energy, Jan. 1986), ch. 6, p. 7.

37 Ibid., ch. 4, p. 5.

38 U.S. Environmental Protection Agency, Standards for Remedial Actions at Inactive Uranium Processing Sites, 48 *Federal Register*, No. 4, 591, (Jan. 5, 1983). The regulation speaks of 0.02 working levels and 0.03 working levels of radon decay products. Since the radiation is commonly presented to the American public in terms of picocuries of radon gas per liter of air, this book uses the picocurie equivalency to working levels whenever practicable.

39 As far back as 1980, David Rosenbaum, deputy administrator for EPA's radiation programs, called indoor radon "the highest radiation danger that the American public faces." Cited in Lisa B. Belkin, "Warning: Home Energy Conservation May Be Dangerous to Your Health," *National Journal*, 2 Aug. 1980, 1274.

40 Refers to citation in note [38] (48 *Federal Register*, 592).

41 *A Citizen's Guide to Radon* (1986 ed.), 2.

42 House Subcomm. on Natural Resources, Agricultural Research and Environment of the Comm. on Science and Technology, *Hearing on Radon and Indoor Air Pollution*, Oct. 10, 1985 (Washington, DC: Government Printing Office, 1986), 90 (testimony by Stanley J. Watras).

43 Ibid., 92.

44 Ibid., 93.

45 Ibid., 115 (testimony by Sheldon Meyers, acting director of EPA's Office of Radiation Programs); and ibid., 134–35 (testimony by John P. Millhone, director of DOE's Office of Buildings and Community Systems).

46 *Reporting on Radon: A Journalist's Guide to Covering the Nation's Second-Leading Cause of Lung Cancer* (Washington, DC: Environmental Health Center of the National Safety Council, Oct. 1989), 14.

47 The EPA's 10% figure contrasted with findings of 6% or 7% by others. Bernard L. Cohen, "A National Survey of Radon–222 in U.S. Homes and Correlating Factors," *Health Physics*, Vol. 51, No. 2 (Aug. 1986), 175–183; A.V. Nero et al., "Distribution of Airborne Radon-222 Concentrations in U.S. Homes," *Science*, Vol. 234, No. 4779 (Nov. 21, 1986), 992–97.

48 *Newsweek*, 18 Aug. 1986, 60–61.

49 U.S. Environmental Protection Agency, *Radon Reduction Methods: A Homeowners Guide* (Washington, DC: U.S. Environmental Protection Agency, Aug. 1986), 1.

50 *A Citizen's Guide to Radon* (1986 ed.), 1.

51 Ibid., 12.

52 Radon Gas and Indoor Air Quality Research Act of 1986, Title IV, Pub. L. No. 99-499, §403, Oct. 17, 1986.

53 Indoor Radon Abatement Act, Title III, Oct. 28, 1988.

54 *Reporting on Radon*, 1.

55 Ibid., 7–8.

56 Ibid., 49.

57 *Comparative Dosimetry in Mines and Homes*, 4.

58 Ibid., 40.

59 Public Service News, The Advertising Council, Inc., New York, NY (circa 1989).

60 Interview, Feb. 12, 1992.

61 F. Henry Habicht, Deputy Administrator, Environmental Protection Agency, Keynote Address, *The 1992 International Radon Symposium on Radon and Radon Reduction Technology*, Minneapolis, MN, Sept. 22, 1992.

62 *A Citizen's Guide to Radon* (1986 ed.), 8.

63 *Hearing on Pending Radon and Indoor Air Legislation*, 4 (statement by Michael Shapiro). In an interview on July 9, 1992, Margo Oge said the "serious message" of the original campaign was replaced in the fall of 1991 by a "humorous" approach. Television announcements showed people engaged in silly activities, while a voice asked the viewer: "What are you doing this weekend that's so important you can't take a little time to test your home for radon, the second leading cause of lung cancer?" A third wave of television advertising planned for 1993 emphasizes the need to "protect your family" and shows a boy and his dog in gas masks. (Copies of the pictures are in Appendix C.)

64 U.S. Environmental Protection Agency, *A Citizen's Guide to Radon: The Guide to Protecting Yourself and Your Family from Radon*, 2d ed. (Washington, DC: Government Printing Office, May 1992). The 1986 *Citizen's Guide* also portrayed dots in a person's respiratory system, but the picture was smaller and without the color of the 1992 version.

65 Ibid., 14. Elsewhere, the EPA frankly stated the reason for its reluctance to acknowledge uncertainty about the basis for its radon policy: "One type of information that does not appear to be effective in educating citizens or persuading them to act is information on uncertainty." The agency supported this notion by citing findings by others that "people dislike uncertainty and may use it as an excuse for disregarding a radon message" or "radon communication should provide only that information most likely to persuade people to take action." Radon Division, Office of Radiation Programs, U.S. Environmental Protection Agency, *Technical Support Document for the 1992 Citizen's Guide to Radon* (Washington, DC: U.S. Environmental Protection Agency, May 1992), ch. 6, 6. Excerpts from the document appear in Appendix D.

66 U.S. Department of Energy, *Radon: Radon Research Program, FY-1990*, (Washington, DC: U.S. Dept. of Energy, March 1991), 247–77.

67 Ibid., 1.

68 Ibid., 283.

69 Ibid., 1.

70 U.S. Department of Energy, Office of Health and Environmental Research, and Commission of European Communities, Radiation Protection Programme, *International Workshop on Residential Radon Epidemiology* (Washington, DC, U.S. Dept. of Energy, July 1989), foreword.

71 Ibid., 18–19.

72 The DOE has been judged far more harshly about policies unrelated to radon, particularly those concerning nuclear power and nuclear waste disposal. The department's reputation has been adversely affected, as Michael Kraft has written, because of its "perceived failure over time to deal openly and competently" with nuclear issues. Michael E. Kraft, "Evaluating Technology through Public Participation: The Nuclear Waste Disposal Controversy," in *Technology and Politics*, eds., Michael E. Kraft and Norman J. Vig (Durham, NC: Duke University Press, 1988), 257.

CHAPTER 2

The Science of Uncertainty

"THERE CAN BE NO SCIENTIFICALLY OR EMPIRICALLY neutral system of language or concepts," wrote Thomas Kuhn in his influential treatise on *The Structure of Scientific Revolutions*.[1] Paul Feyerabend holds that scientific truth is determined by scientists and politicians, based on money, interest, and pride.[2] Langdon Winner recounts that even artifacts born of science and technology may be political insofar as they "contain possibilities for many different ways of ordering human activity."[3]

Politics doubtless informs the ordering of human activity on the indoor radon issue, as it does all matters of public policy. Nevertheless, whether politically configured or not, some scientific facts are universally accepted. In a philosophical sense, of course, every scientific fact remains uncertain since it is subject to testing and reinterpretation. But in practice, explanations of phenomena are considered certain when the scientific community is in agreement derived from knowledge at a particular time.

This chapter explores the elusive interface between scientific certainty and uncertainty in the context of the radon issue. The nexus of certainty and uncertainty affects the central issue of the book: What should be done about indoor radon as a matter of public policy. The most important question in this regard is whether a substantial number of people are endangered by indoor radon. If they are, where and to what extent is this the case, and what can be done to lessen the risks? The questions relate to understandings about radiation.

Radiation, Radon, and Uncertainty

Some characteristics of radon are beyond dispute: its atomic structure, its place on the radioactive decay chain, its omnipresence, the
radioactive nature of its short-lived progeny. Moreover, no one doubts
that exposure to high concentrations of radon progeny can be dangerous to human health.

But deciding exactly how much radiation is necessary to cause
damage eludes scientific consensus. If a radiation dose is low, the
body can usually repair or replace damaged cells without detectable
health effects.[4] Many scientists adopt a conservative approach, however, and assume that *any* amount of radiation should be considered
harmful. Thus, the degree of presumed danger from low amounts of
radiation is extrapolated from confirmed dangers of high levels. Even
at high concentrations, however, linear relationships are not always
demonstrable. The most extensive investigations of the subject involved the tens of thousands of Japanese survivors of the atomic
bombings in World War II.

Hundreds of studies of these people were reviewed by The Committee for the Compilation of Materials on Damage Caused by the
Atomic Bombs in Hiroshima and Nagasaki. Exposure to high levels of
radiation was found to have increased the incidence of some types of
cancer, but not all. The clearest relationship was found between exposure and leukemia, where a linear association was evident among victims
in both Hiroshima and Nagasaki.[5]

Despite a demonstrated linear relationship at high levels of radiation, the evidence was contradictory at exposures to low concentrations. At levels below 100 rads, for example, an effect was observed in
Hiroshima, but none in Nagasaki.[6] (A rad, which stands for "radiation
absorbed dose," is a measure of the amount of absorbed radiation
energy by the tissues of the body.) The committee conjectured that
the difference might have arisen from types of rays that predominated
in each experience. The nuclear components of the two bombs were
different, and consequently in Hiroshima there were thought to be
more neutron rays emitted relative to gamma rays, while in Nagasaki
the reverse was the case.[7] Subsequent calculations suggested that
neutron exposure in Hiroshima was lower than originally estimated
and may have had little bearing on the matter.[8]

Whatever the explanation, the contrasting experiences with leukemia in the two cities demonstrate only one of many enigmas about
the effects of low-level radiation. When radiation from the atomic
bombs was measured against the incidence of lung cancer, no association whatsoever was found at lower radiation levels. Only when exposure

was 200 rads or greater were more lung cancers detected.[9] Thus, in some cases a threshold may exist below which harm from radiation does not occur. The matter remains uncertain because the findings may have resulted from inadequate statistical power. If many more people had been exposed, perhaps then a disproportionate increase in the number of lung cancers could have been detected.

The relevance of the linear-versus-threshold argument is central to the indoor radon debate. Increased exposure to high levels of radon progeny unarguably increases the risk of lung cancer. But unlike the gamma rays from the atomic bombs, radiation from the progeny that causes lung cancer is in the form of alpha particles. While gamma rays pass through human tissue, the more massive alpha particles do not. They are largely stopped by the skin where their energy is quickly dissipated without causing damage. But if inhaled, radon progeny may reach cells of the respiratory tract, which are more sensitive to the ionizing changes that can lead to cancer. Basal cells lie near the surface of the lung lining, and because they are continuously dividing, they are particularly susceptible to the effects of radiation.

The information about this process and the relationship of radon progeny to lung cancer has been developed from investigations of miners and of experimental animals. All models about the presumed effects of radon on home dwellers derive from these studies.

Miner Studies

The two most comprehensive assessments of the relationship of radon to lung cancer are in Report No. 78 published in 1984 by the National Council of Radiation Protection and Measurements (NCRP), and the BEIR IV report published in 1988 by the National Research Council of the National Academy of Sciences.[10] Both offer extensive reviews of the studies of miners, particularly uranium miners, that demonstrate the effect of radon on human health. The NCRP report cited results from miner studies in the United States, Czechoslovakia, Sweden, Canada, Great Britain, and France. All showed substantially elevated incidences of lung cancer among miners.[11]

BEIR IV focused on the cohorts of four studies—uranium miners in Ontario, Saskatchewan, and the Colorado Plateau (in Colorado and Utah), and metal miners in Sweden. A review of the Colorado Plateau miners between 1959 and 1977 revealed numbers of lung cancer deaths as much as six times higher than would be expected in the general population.[12]

One of the committee members who wrote the BEIR IV report was Jonathan Samet of the University of New Mexico School of

Medicine. Samet subsequently coauthored with Richard Hornung of the National Institute for Occupational Safety and Health a review of several miner studies and risk assessment models. The review included five representative miner populations from the 20 or so that have been investigated over the years and found "several consistent patterns." The most compelling was the increased incidence of lung cancer among the miners, evidently as a consequence of their exposure to radon decay products. Despite differences in analytical techniques and exposure data among the studies, the risk estimates were "remarkably homogeneous" at about 1.5 percent per working level month (WLM) exposure.[13] (A WLM is exposure to one working level (WL) for 170 hours. Thus, one WL would be the decay products in equilibrium with 300 picocuries of radon gas per liter of air. This assumes an equilibrium factor of 30 percent, which is now common in ventilated mines.[14])

The review noted that beside cumulative exposure, the studies pointed to other influences on the risk of lung cancer. Risk is higher among smokers, for example, and it is greater if exposure is at an older age. The studies also suggested a greater hazard with lower exposure rates for longer duration than higher exposure rates for shorter duration, when cumulative exposure was equal.[15]

Uncertainties in Extrapolations

The findings among miners were in the realm of very high exposure levels, where radon concentrations typically ranged in the hundreds or thousands of picocuries per liter of air. As is common among all miner studies, the implications for lower-level exposures are left unclear. Samet and Hornung mention the dearth of evidence concerning lower levels as one of several uncertainties about the radon issue.

> The quantitative relationship between exposure to radon and radon decay products and lung cancer risk has not been precisely described, and uncertainties about the effects of age, gender, cigarette smoking, and other factors on this relationship await resolution. Extrapolation of risk estimates based on studies of miners to the general public requires assumptions in areas of uncertainty. We also lack exposure information based on a large and representative sample of the nation's homes.[16]

These doubts are underscored in the NCRP and BEIR reports, as they are in other studies that propose models to estimate risks at lower levels of radon. In reviewing the miner investigations, NCRP Report No. 78 concluded that "[t]here are many factors that make it difficult to interpret the human epidemiological data with great accuracy." The list of factors is imposing. It includes gaps in past studies about cumulative exposures and imprecision about the ages of people when they were first exposed. Some studies counted only deaths caused by cancer, while others included living cancer cases. Cigarette consumption was not always taken into account, and follow-up time among the studies varied.[17] Despite these uncertainties, the NCRP developed a model of risk estimates at lower radon concentrations, allowing that it was based on assumptions rather than confirmed information.

Added to the uncertainties born of incomplete data were those built into the model. The NCRP model assumed, for example, that smoking had an additive rather than multiplicative effect. That is, the excess risk from radon was added to the background rates for smokers and nonsmokers. A multiplicative, or synergistic, interaction assumes that the risk for smokers is multiplied by the risk for radon exposure. This would presume that the risks from smoking and radon are greater when the two are in combination than when taken separately.[18]

Other investigations that assumed a synergistic relationship, including the BEIR IV report, accordingly contained higher risk estimates. BEIR IV also acknowledged other lapses in information. The BEIR IV calculations were based on four miner studies that offered uncertain estimates in exposures as well as in classification of disease. But the most critical gap appears to be in the area of smoking. Nonsmokers, BEIR IV recognized, "were poorly represented in the lung-cancer mortality data" available from the studies, and information about individual smoking habits was available only for the Colorado miners.[19] Thus, three of the four miner studies used as the basis for BEIR IV's epidemiological evidence provide little help with the most important confounding issue: the relationship between smoking and radon progeny as causes of lung cancer.

The uncertainties in model building are exemplified by the large disparity between estimates provided by the various reports. The BEIR IV risk estimate, for example, was about three times greater than that of the NCRP study. BEIR IV proposed that the lifetime risk of dying from lung cancer due to lifetime exposure to radon progeny is 350 per million-person-WLM. (This, according to BEIR IV, is equivalent to saying that if one million people were exposed to 4 picocuries of radon per liter of air for 1 year, 350 would be expected to die from lung cancer in the course of a lifetime.) The NCRP figure, as the BEIR IV authors noted, was 130 per million-person-WLM.[20]

Most striking, however, is the recognition in virtually every scientific report on the subject that the effects of radon levels commonly found indoors are speculative. An assessment of lung cancer risks from indoor radon by the International Commission on Radiological Protection (ICRP) concluded that "further investigations are necessary to confirm, or to improve, this risk assessment."[21] A standard recommended by the National Institute for Occupational Safety and Health is based on "uncertainties [that] include the choice of risk assessment method and model, the measurement methods used for data collection, and risk estimates derived from data that are heavily weighted with higher exposures."[22] According to the NCRP report:

> None of the studies, so far, has produced data which show a statistically significant excess of lung cancer in the lowest cumulative exposure category (less than 60 WLM) [which is equivalent to exposure for one year to residential concentrations of 240 picocuries of radon per liter of air]. Therefore, in estimating the effect of radon daughter exposure at environmental levels, normally less than about 20 WLM [or one year's exposure to 160 picocuries] per lifetime, the attributable risk at high exposures must be extrapolated to the low exposure region.[23]

The wording is convoluted, but the message is clear: The NCRP report says that we do not know if radon at levels typically found indoors has any effect on people's health. Similarly, the BEIR IV committee recognized the following:

> Several assumptions are required to transfer risk estimates from an occupational setting to the indoor domestic environment. Accordingly, the committee assumed that epidemiological findings in the underground miners could be extended across the entire life span, that cigarette smoking and exposure to radon daughters interact multiplicatively, that exposure to radon progeny increases the risk of lung cancer in proportion to the sex-specific ambient risk of lung cancer associated with other causes, and that, to a reasonable approximation, a WLM yields an equivalent dose to the bronchial epithelium in both occupational and environmental settings.[24]

A subsequent analysis by another committee of the National Research Council elaborated on the relationship of radon exposure in mines to homes. It examined the assumption made in previous studies that the effects of residential radon may be extrapolated directly from those in the mine. The committee developed a model indicating that equivalent amounts of alpha radiation in the two environments would affect miners more than home dwellers. It estimated that adults at home would have 30 percent less radiation exposure to their lungs than would miners. In presenting these findings the committee acknowledged that its model suffered from uncertainties common to other investigations—for example, about breathing rates among working miners and the effects of aerosol particles in mining environments. Indeed, the report acknowledged that "any dosimetric model, regardless of its sophistication, inevitably simplifies extremely complex physical and biological phenomena."[25]

The Mining Environment

The differences between environments in the mine and home seem inadequately appreciated by many investigators. Congressional testimony in 1982 by men who had worked in uranium mines in the 1950s and 1960s is instructive. Leo Redhouse, Sr., a Navajo Indian, told of his experiences about working in the midst of "intense smoke." "The mines I worked in were all unsafe, with very little ventilation.... We were told to go back into the mines 5-10 minutes after each dynamite blast."[26] Sam Jones, another Indian miner, recalled that "[a]fter dynamite blasting, the mines were choked with dust. We had to set up ventilation equipment ourselves. We ate our lunches down in improvised so-called lunch rooms inside the mines to save time. Drinking water was made available to us inside the mines. Smoke was intense from the mining machineries."[27]

Graphic descriptions like these are absent from scientific reports. Although anecdotal, they suggest incalculable differences between home and mining environments, especially before ventilation in mines was installed during the 1970s. To assume that a direct extrapolation can be made without accounting for the differences in the environments seems unrealistic. To be sure, many studies recognize that extrapolations from mining to residential experience are speculative. Every respected scientific review of the risk estimates of indoor radon acknowledges uncertainties. Invariably, the studies call for further investigation to confirm whether their speculations are valid.

Animal Studies

The other principal scientific basis for interpreting the radon-cancer connection is from laboratory studies involving animals. The NCRP and BEIR IV reports offer overviews of investigations that seek a relationship between radon progeny and lung cancer in laboratory mice, rats, hamsters, and dogs. Although not all investigations in which animals were exposed to elevated radon levels resulted in lung cancers, the NCRP found that as a whole the animal experiments "suggest that exposure rate influences tumor production."[28] At the same time, NCRP indicated that "different sites [are] affected in animal lung versus human lung."[29]

BEIR IV noted that lung cancers have been induced in rats at relatively low exposures, around 20 WLM. (This is equivalent to 80 picocuries of radon per liter of air for one year, which is the presumed average lifetime exposure of Americans in their homes.) Work with dogs has been in the hundreds of WLM range. Moreover, the relationship between animal cancers and radon, whatever the level, leaves uncertain its applicability to humans. BEIR IV observed that "the location and histopathology of such cancers are not analogous to humans, and caution is warranted in extrapolating from experiments with laboratory animals to humans."[30]

Before exploring the implications of these findings for public policy, there are two other scientific-technical questions associated with radon policy to be reviewed: determining which homes and other structures contain supposedly unsafe levels of radon, and how they may be mitigated. Measurement and mitigation techniques help order the possibilities of radon policy. They are, as Langdon Winner might say, the artifacts of radon science and technology.

Radon Testing

At first glance, establishing radon concentration in an enclosed structure seems uncomplicated. The quickest method is "grab-sampling," which involves isolating a small volume of air and measuring. An electronic radon detector can provide an almost immediate reading of radon and radon daughter concentrations in an air sample. Sophisticated detectors can cost several thousand dollars, however, and are used infrequently.

The most commonly used short-term measuring device is the charcoal canister. Costing only a few dollars, a small canister of activated charcoal is set in place for 2 to 7 days, then sealed, and sent to a laboratory for analysis. There, a gamma-ray detector measures the amount of radiation being emitted by the charcoal. (The quantity of gamma radiation emitted during radon decay is minimal but detectable.) The amount of radon in the charcoal can thus be estimated, which is a reflection of the amount in the air at the time of exposure and the concurrent concentration of radon daughters.

In some communities, the short-term charcoal test has become a virtual requirement for real estate transactions. Yet on two counts the technique is subject to question. The first involves accuracy of measurement. While some states have begun to monitor and certify radon technicians and techniques, even when reputable firms are used the results can be problematic. A New Jersey study, for example, revealed serious discrepancies in test results by laboratories that had been approved by the state's Department of Environmental Protection. Three charcoal canisters exposed at the same time in one room gave readings ranging from 6 to 14 picocuries per liter of air. The reasons for the discrepancies were unclear. They could have occurred because of lags between time of exposure and laboratory analysis, movement of canisters during exposure period, or errors in measurement.[31] Nevertheless, when conducted with care and competency, accurate measurements are technically possible.

More problematic is whether any short-term measurement can be a proper indicator of risk. Not only do concentrations of the gas often differ in one part of a house from another, they vary according to seasons and weather conditions. Outdoor measurements taken throughout the year at single locations have ranged between 0.1 and 0.4 picocuries per liter of air.[32] Indoor variations may be far greater. Studies have found, for example, that some homes with 10-picocurie per liter levels in the winter showed only one picocurie in the summer.[33] David J. Brenner, a radiation oncologist at Columbia University, cites tests in the living room of a house where radon levels varied 23-fold during a 10-day period. One measurement yielded a result of 6 picocuries per liter of air, but 3 days later a measurement in the same location showed 140 picocuries.[34]

In a year-long study, Naomi Harley, a radiation physicist at the New York University School of Medicine, detected a 30-fold difference in the readings of a home that "was constructed to be as 'radon resistant' as possible." Built in 1988, continuous radon monitoring during the following year revealed that 2-day measurements in the basement varied between 0.49 and 14 picocuries.[35] These and similar findings have led to questions about the value of short-term testing for any purpose having to do with safety or risk. In Finland, as described in Chapter 9, short-term testing is deemed worthless; the authorities require that measurements be taken for at least 2 months.

The EPA's 1986 brochure, *A Citizen's Guide to Radon*, exemplifies the dilemma. While encouraging homeowners to have a short-term test, the brochure acknowledged that such measurement is not a reliable indicator of the average radon level. Rather, measuring for a few days will indicate "the *potential* for a radon problem" (original emphasis) that should determine whether follow-up testing is advisable.[36] The 1992 *Citizen's Guide* somewhat altered the perspective and said that "a short-term test followed by a second short-term test may be used to decide whether to fix your home."[37]

But as indicated, a short-term reading, even if accurate for the brief period in question, can be very misleading. A reading might show less than 4 picocuries, obscuring the possibility that year-long average levels could be much higher. If EPA's advice were followed, a homeowner would never find this out.

The most commonly used longer term testing mechanism is called an alpha track detector. Its main component is a strip of plastic sealed in an air-tight can about two inches in diameter. When the can is unsealed and exposed to the air, alpha particles create a microscopic etch or track on the plastic as they contact it. After a specified period, the can is resealed and sent to a laboratory where the plastic is chemically treated and the tracks are counted. The number provides an estimate of the radon concentration during the time of exposure.

The 1986 *Citizen's Guide* suggested that the alpha track detector be used for at least 2 to 4 weeks. Although the *Guide* refers to the period as a minimum, the implication is that a few weeks would be adequate. But measurements taken for greater periods have proved problematic. Harley found that even 90-day average measurements could be misleading.[38] The average radon level in a house she studied for a year was lower than 4 picocuries per liter of air. But in assessing 12 90-day measurement periods during that year, for two of these periods, or 17 percent of the time, average radon readings exceeded the 4-picocurie action level. Presumably, if a homeowner heeded results from either of these 3-month testing periods, he should have had his house remediated. He would not have realized that the annual average was well below the 4-picocurie level and that mitigation was unnecessary even from the EPA's perspective. Moreover, remediation may have its own problems.

Remediation

During the past few years, millions of dollars have been spent in state and federal programs to develop techniques to reduce indoor radon concentrations. Although mitigation technology is imperfect, and for some homes remains a matter of trial and error, capability in this area has improved substantially. A 1986 report prepared for the Department of Energy, and published by the National Technical Information Service (NTIS), reviewed techniques to reduce the concentration of radon, including removal of its source, isolating the source, and air ventilation.[39]

The report noted that not all techniques may be appropriate to all homes. To remove radioactive soil from under an existing structure could be prohibitively expensive. In Montclair, New Jersey, the cost for removal and refurbishing had originally been figured at $600,000 for each home. (As discussed in Chapter 6, this was before lawsuits and when finding a depository for the soil drove the expenses far higher.) When the design or content of structural materials like cement blocks are the source of radon, removal might destroy the building.

As mentioned in Chapter 1, the cost of attempted remediation for each home in Grand Junction, Colorado, averaged about $13,500 in the 1970s, according to the NTIS report. Before cleanup, 71 percent of the houses had radon levels above 6 picocuries per liter. After cleanup, though some homes with very high readings showed considerable drops, 68 percent still had readings exceeding 6 picocuries.[40] Thus, the result of even this ambitious effort, where the offending soil was dug up and carted away, was ambiguous. The cost was substantial, and the measure of "success"—6 picocuries—is higher than the EPA's acceptable number.

Other techniques also have problems. Sealants that are intended to isolate radon-producing material can deteriorate and, therefore, require continued monitoring. Harvey Keaton, of the Florida Office of Radon Control, has pointed out that there is "no way that you can have a little radon seep in and then not have a lot as it accumulates."[41] (Actually, if the rate of entry is low, the indoor concentration will be low; if high, so will the indoor concentration be high. To be effective, the sealing job must be perfect.)

As for ventilation, attempts to enhance air exchange have shown some unwelcome consequences. Although ventilation can reduce radon levels, the NTIS review indicates that it sometimes has the opposite effect. Inappropriately designed ventilation may lower indoor air pressure and "actually pull in more radon than is expelled."[42]

Despite these uncertainties, by the early 1990s experience with mitigation efforts in thousands of houses had enhanced the body of technical knowledge. The EPA was indicating that most homes could be remediated for an initial cost of about $1,000 or $2,000. Monitoring and maintenance would be necessary forever after, however, and would require additional expenditures.

The most common mitigation arrangement involves "sub-slab" ventilation. This usually involves installation of pipes below the basement floor. These are connected to one or two vertical pipes that extend along the walls of a house and open into the outside air. Suction fans near the tops of the extension pipes draw the sub-basement air through the piping system for release outdoors. The EPA has found that this technique can usually reduce inside radon levels to below 4 picocuries per liter of air.[43] The technique would be inapplicable or prohibitively expensive in some circumstances—for instance, when cement blocks in a structure are the main cause of elevated radon, or when a home is built on a rock foundation that precludes easy manipulation below its lowest floor.

Moreover, while technical experts generally regard sub-slab ventilation as the most successful mitigation approach, it has pitfalls. John Spears, an architect with the National Association of Home Builders who has worked with government agencies on questions of home design and radon levels, expresses concern. He wonders about the long-term reliability of systems that depend on fans or other active parts. "Talk to me about these houses a few years from now. Take the measurements then. See if the fans are working when new owners have moved in. You know, there are no automatic signals [in many systems] to indicate when the fan motors stop or burn out."[44]

In 1992, Kenneth Wiggers, owner of a mitigation systems company in Iowa, introduced a new dimension of concern. During the first half of the year he stopped installing sub-slab ventilation equipment after discovering that it was drawing air out of homes. He acted after residents of homes in which systems were placed began to complain of lethargy and short-term memory loss. The radon mitigation systems evidently were causing backdrafts, drawing gases from furnaces, water heaters, and dryers into the home. In a paper titled "Cold Climate Radon Mitigation Systems: Opportunities and Perils," he described oversized systems as "likely to be an immediate threat to the health and safety of people."[45]

Thus, while radon measurement and remediation are technically possible, their operations may be more complex than has been presented to the public. In the EPA's effort to encourage people to become concerned about radon, and to do something about levels above 4 picocuries per liter of air, the benefits of testing and fixing have been made to appear more certain than seems warranted.

Epidemiology and Indoor Radon

The range of uncertainty about potential radon-caused deaths is dealt with in an edited volume on *Indoor Radon and Its Hazards*. The lead author, David Bodansky, noted that the number of deaths caused by radon exposure cannot be known with precision. The estimate, he says, could run from near zero to 50,000. Nevertheless, "[i]t is considered prudent for purposes of radiation protection to assume that linearity holds," said Bodansky, and "that the United States cancer toll from radon is about 10,000 deaths per year."[46]

But Bodansky exemplifies the ambiguity of the situation in two subsequent passages. He noted that "the guiding principles for the regulation of radon exposure are that there is no level of exposure below which one can be certain that there will be no adverse health effects when a large population is exposed."[47] At another point, however, he acknowledged that "the absence of compelling direct evidence of increased lung cancer rates in regions of high indoor radon concentration raises doubts about the validity of the linearity hypothesis."[48]

In recent years several empirical studies have been undertaken to investigate the relationship of radon to the incidence of lung cancer in the general citizenry. Many remain in progress, but those completed show little evidence of a correlation. This may be because there is no relationship. But it also could be attributable to the design of the studies or to insufficient numbers of cases to provide adequate statistical power. Thus, the lack of correlation reported by Ralph Lapp and Bernard Cohen was unenthusiastically received by several epidemiologists.

Lapp, a physicist and radiation safety consultant, observed that natural radon levels in New Jersey are seven times greater than in Texas, (5.4 picocuries compared to 0.8). Yet lung cancer deaths as a ratio of total cancer deaths in the two states are almost the same (26 percent in New Jersey, 29 percent in Texas).[49] Cohen's measurements in 39,000 homes in 415 counties actually showed an inverse relationship. Cohen, a University of Pittsburgh physicist, found a strong tendency for counties with high radon levels to have low lung-cancer rates.[50]

Susan Conrath of the EPA's radon division and other epidemiologists dismiss these as ecological studies that suffer from incomplete designs. Because the group, rather than the individual, is the unit of analysis, the ecological approach is "incapable of assessing individual smoking experience or the effects of family mobility, both very significant variables in radon risk assessment."[51] Smoking, length of residency, and other variables are undoubtedly important. But even in studies where smoking and mobility are accounted for, a connection between low-level indoor radon and the incidence of lung cancer has not been clearly established.

Studies of populations in China seem responsive in part to these methodological objections. One included 73,000 inhabitants of regions with high-background radiation, compared with 77,000 inhabitants of nearby low-background regions. Despite the fact that more than 90 percent of the families of the inhabitants in the high-background areas lived there for six or more generations, no significant differences in lung-cancer rates between the groups could be found.[52]

A study of women in the industrial city of Shenyang showed

similar results. Year-long radon measurements were taken in the homes of 308 women with newly diagnosed lung cancers, and in the homes of 356 control subjects. Interviews established the smoking status of the subjects and the length of residence—the median time was 24 years. The measurements showed the median household radon level to be 2.3 picocuries per liter of air, with 20 percent exceeding 4 picocuries. Yet no statistically significant association was observed between radon and lung cancer.[53]

The Shenyang investigation was not an ecological study, but rather a case-control study. In case-control investigations a group of cases is compared with a group of noncases (controls), and Conrath calls this the "design of choice for assessing radon health risk." The report was published after Conrath's article appeared, so she might not have known about its findings. The only case-control study she mentioned in her article is a New Jersey investigation that, she said, showed a "borderline significance level" of increasing risk with increasing residential radon concentrations.[54] Discussed in Chapter 8 are the New Jersey findings and the fact that attributing significance to them, "borderline" or otherwise, is of doubtful warrant.

Meanwhile, in 1989 and 1991 the Department of Energy and the Commission of European Communities sponsored international workshops on "Residential Radon Epidemiology." All known investigators with active projects on the subject were invited. Most of the studies were still in progress, and results were not available. But investigators outlined the projects under way in 10 countries: the United States, Canada, Norway, Sweden, Finland, the United Kingdom, Belgium, Germany (Federal Republic), France, and the People's Republic of China.[55]

As of 1992, four of the studies were published either as complete or preliminary reports. None was able to identify an unambiguous association of lung cancer risk with radon levels in the home. The New Jersey study noted a statistically significant "trend" for increasing risk with increasing radon exposure, but at the same time found that "the relative risk estimate was not statistically significant."[56] The Shenyang, China, study found that "[n]o association between radon and lung cancer was observed regardless of cigarette-smoking status, except for a nonsignificant trend among heavy smokers."[57] The study in Finland concluded: "The present study did not establish the anticipated association between indoor radon exposure and the risk of lung cancer in Finland."[58]

A report of the Swedish study said that its "risk estimates appeared within the same range as those projected for miners."[59] One of the authors, John Boice, an epidemiologist with the U.S. National Cancer Institute, characterized as "weakly positive" the relationship found in the study between lung cancer victims and residential radon. But he noted that "when adjusting for length of time of occupancy, then the results disappear altogether—the dose-response relationship goes away." Further, he said, "If you use the National Academy of Science's BEIR IV model with the Swedish data, you have no relationship at all."[60]

Despite the failure thus far to confirm a relationship between lung cancer and low-level radon exposure, many epidemiologists maintain that the risk model is not necessarily invalid. Rather, the number of cases in each study may be too small to disconfirm a linear effect. The most that can be said is that the results are consistent either with the linear hypothesis *or* with no effect from radon. An article by Jay Lubin, Jonathan Samet, and Clarice Weinberg made precisely this point.[61]

The authors held that a larger number of subjects than now commonly used in case-control studies would be required to "substantially reduce the uncertainty of extrapolating from mines to homes." They concluded that because of modestly sized cohorts in current studies, "predictably" some will not show increased risk from indoor radon, while others may show effects "substantially above levels expected based on miner studies."[62]

Samet was a cochairman of the Workshops on Residential Radon Epidemiology sponsored by the Department of Energy and the Commission of European Communities in 1989 and 1991. The other cochairman was Jan Stolwijk, professor of epidemiology at Yale University School of Medicine. Stolwijk is currently collaborating on a radon case-control study in Connecticut and is familiar with the other epidemiological radon studies published or in progress. His assessment: "No reputable epidemiologist is impressed with current findings linking indoor radon to lung cancer."[63]

Stolwijk, like many epidemiologists, does not reject the possibility that linkage exists. Rather, he emphasizes, the matter is uncertain.

Health Physicists

In the late 1980s, debate about the wisdom of the EPA's radon policy remained largely out of public view. Within the health physics community, however, it became intense. Comprised of more than 6,000 members, the Health Physics Society is dedicated, in its own words, to "the development of scientific knowledge and practical means for the protection of man and his environment from the harmful effects of radiation, thus providing for its utilization for the benefit of mankind."[64]

The society publishes the scientific journal *Health Physics* and a monthly newsletter that contains scientific articles, comments, reviews, and general information. By the end of the decade, few editions of the journal or newsletter failed to carry items about indoor radon. The January 1991 newsletter dealt almost exclusively with the matter, the first time an edition was primarily devoted to a single subject. Insofar as the society houses the largest aggregation of informed professionals about health physics in the United States, the organization's considerations are particularly instructive. Indeed, the articles and comments about radon included forceful advocacy of all positions.

In an article titled "If Indoor Radon Isn't a Problem, Then There Are No Problems," Daniel J. Strom of the University of Pittsburgh warned: "Not everyone is at high risk, but a sizeable number of people are."[65] JoAnne D. Martin of Douglas Martin and Associates advocated a dramatic education effort to alert the public to the risk. "If the information is shocking, if it makes people feel uncomfortable, so much the better. A spark of controversy may wake people up to this serious issue."[66]

Conversely, the newsletter published a statement by an Ad Hoc Working Group on Radon that rejected such notions. In 1988, Ronald Kathren, then president of the society, established the five-member working group to draft a position paper on cancer risk and indoor radon. Chaired by William A. Mills, it produced a paper whose tenor is revealed in the opening paragraph.

> Inadequate information on radon health risks and the meaning of screening measurements is leading many homeowners to spend money on reducing indoor radon that may not significantly reduce their risk of lung cancer. The widely publicized estimate of 20,000 lung cancer deaths a year due to indoor radon implies that reductions in radon could save 20,000 lives a year. This will not happen because more than 70 percent of those estimated deaths are due to the combined effects of radon and cigarette smoking. Homeowners should understand that reduction in radon levels alone may not reduce the total number of lung cancer deaths in the United States. Public officials should know that the appearance of a national radon problem is greatly exaggerated when EPA's screening measurements are used to assess the extent of the problem.[67]

Although the mandate to the working group was to produce a statement that would represent the society's position, the organization's Scientific and Public Issues Committee decided the working group's draft was too contentious. The committee then formulated one that the society's president, Genevieve Roessler, released in 1990 as the Health Physics Society's official "Position Statement: Perspectives and Recommendation on Indoor Radon."

Less harsh than the working group's draft, the final statement was nevertheless critical of existing national policy. The statement encouraged "public understanding of the potential risks of radon" and recommended that exposure to the gas "be minimized in accordance with practical considerations." It then offered seven observations and recommendations that amounted to a tempered rebuke of the EPA. They included recommendations that the agency "review its emphasis on the use of 4 picocuries of radon per liter of air (pCi/L) as an action level. Rather, the EPA should emphasize the prompt identification of indoor occupied areas with very high radon concentrations (i.e., tens of pCi/L and greater) as candidates for prompt mitigation."

While the Health Physics Society endorsed efforts to "provide realistic information on the potential benefits of radon reduction in homes," it urged that education "should be based on reason, rather than emotion." Its concluding observation was a reminder that "[a]lthough we know a great deal about radon and its potential effects on health, there is still much more we do not know and could benefit from learning."[68] (Full text of the statement appears in Appendix E.)

In an unusual action, accompanying the position statement in the January 1991 newsletter was a questionnaire soliciting the views of the society's members about its contents. The questionnaire offered a scale of choices about each section of the statement: totally agree, generally agree, neutral, generally disagree, totally disagree. In the August newsletter, Roessler reported the results. The 102 members who responded overwhelmingly endorsed the introductory passage and all seven sections: The sum of "totally agree" and "generally agree" responses for each section ranged from 72 percent to 87 percent.[69]

Did the 102 respondents reflect the views of the society's larger membership? Roessler believes they did, based on her personal contacts with many members.[70] Moreover, if members felt the statement varied substantially from their views, more could have been expected to register disagreement.

Other prestigious societies, including the American Medical Association (see Appendix E) and the American Lung Association, had taken positions endorsing the EPA's policy. But the views of the health physics community—a group intimately involved with the subject—deserve particular respect. As the next two chapters show, however, thoughtful spokespersons represent a variety of positions about radon policy. In systematically reporting their views, while noting their backgrounds and the nature of their expertise, the chapters flesh out the human dimensions that underlie the radon debate.

Notes

1   Thomas S. Kuhn, *The Structure of Scientific Revolutions*, 2d ed. (Chicago: University of Chicago Press, 1970), 146.

2   Paul Feyerabend, *Against Method* (London: NLB, 1976), 188–89, 302–3.

3   Langdon Winner, "Do Artifacts Have Politics?" in *Technology and Politics*, eds., Michael E. Kraft and Norman J. Vig (Durham, NC: Duke University Press, 1988), 42.

4   U.S. Department of Energy, Assistant Secretary for Nuclear Energy, Office of Program Support, *Understanding Radiation*, (Washington, DC: U.S. Dept. of Energy, 1986), 11.

5   The Committee for the Compilation of Material on Damage Caused by the Atomic Bombs in Hiroshima and Nagasaki, *Hiroshima and Nagasaki: The Physical, Medical, and Social Effects of the Atomic Bombings* (New York: Basic Books, Inc., 1981), 252.

6   Ibid., 264.

7   Ibid., 265.

8   William C. Roesch, ed., *U.S.-Japan Joint Reassessment of Atomic Bomb Radiation Dosimetry in Hiroshima and Nagasaki*, Vol. 1 (Washington, DC: Radiation Effects Research Foundation, 1987), 130–38.

9   The Committee for the Compilation of Material on Damage Caused by the Atomic Bombs in Hiroshima and Nagasaki, 289. (Refers to citation in note [5]).

10 NCRP Report No. 78, *Evaluation of Occupational and Environmental Exposures to Radon and Radon Daughters in the United States* (Bethesda, MD: National Council on Radiation Protection and Measurements, 1984); Committee on the Biological Effects of Ionizing Radiations, National Research Council, *Health Risks of Radon and Other Internally Deposited Alpha-Emitters, BEIR IV* (Washington, DC: National Academy Press, 1988).

11 NCRP Report No. 78, 96–108.

12 BEIR IV, 448.

13 Jonathan M. Samet and Richard W. Hornung, "Review of Radon and Lung Cancer Risk," *Risk Analysis*, Vol. 10, No. 1 (1990), 69–70.

14 The 300-picocurie figure is based on calculations by Naomi H. Harley and John H. Harley, "Potential Lung Cancer Risk from Indoor Radon Exposure," *Ca—A Cancer Journal for Clinicians*, Vol. 40, No. 5 (Sept./Oct. 1990), 267.

15 Samet and Hornung, 69–70.

16 Ibid., 73.

17 NCRP Report No. 78, 109.

18 Samet and Hornung, 70.

19 BEIR IV, 43–45.

20 Ibid., 76–77.

21 *ICRP Publication 50: Lung Cancer Risk from Indoor Exposures to Radon Daughters*, A Report of a Task Group of the International Commission on Radiological Protection (New York: Pergamon Press, 1987), 52.

22 U.S. Department of Health and Human Services, National Institute for Occupational Safety and Health, *Criteria for a Recommended Standard to Occupational Exposure to Radon Progeny in Underground Mines* (Cincinnati: NIOSH Publications, Oct. 1987), iv.

23 NCRP Report No. 78, 112.

24 BEIR IV, 7–8.

25 Panel on Dosimetric Assumptions Affecting the Application of Radon Risk Estimates, National Research Council, *Comparative Dosimetry of Radon in Mines and Homes* (Washington, DC: National Academy Press, 1991), 4.

26 Senate Comm. on Labor and Human Resources, *Hearing on Radiation Exposure Compensation Act of 1981—Part 2*, Apr. 8, 1982 (Washington, DC: Government Printing Office, 1982), 131 (statement of Leo Redhouse, Sr.).

27 Ibid., 133 (statement of Sam Jones).

28 NCRP Report No. 78, 148.

29 Ibid., 145.

30 BEIR IV, 441.

31 Bettina Boxall, "8 Tests and 6 Different Measurements," *The Record* (Hackensack, NJ), 19 May 1986, A-6. After finding that some radon-testing firms had an average error rate of nearly 50%, Public Citizen, a consumer group, called for a national quality control and certification program. "Group Says Sloppy Labs Mislead Consumers on Radon Tests," *New York Times*, 5 Jan. 1989, A-16.

32 Anthony V. Nero, Jr., "Radon and Its Decay Products in Indoor Air: An Overview," in *Radon and Its Decay Products in Indoor Air*, William W. Nazaroff and Anthony V. Nero, Jr. (New York: John Wiley and Sons, 1988), 42.

33 Andreas C. George and Lawrence E. Hinchcliffe, "Measurements of Radon Concentrations in Residential Buildings in the Eastern United States," in *Radon and Its Decay Products*, ed., Philip K. Hopke (Washington, DC: American Chemical Society, 1987), 47–57.

34 David J. Brenner, *Radon: Risk and Remedy* (New York: W.H. Freeman and Co., 1989), 115.

35 Naomi H. Harley and Passaporn Chittaporn, "Comparison of Measurement Protocols for Radon in an Ultra-High Energy Efficient Home," presented at a Science Writers Workshop on "Radon Today: The Science and the Politics," sponsored by the U.S. Dept. of Energy in Bethesda, MD, Apr. 25–26, 1991.

36 U.S. Environmental Protection Agency, *A Citizen's Guide to Radon: What It Is and What to Do About It* (Washington, DC: U.S. Environmental Protection Agency, Aug. 1986), 7.

37 U.S Environmental Protection Agency, *A Citizen's Guide to Radon: The Guide to Protecting Yourself and Your Family from Radon*, 2d ed. (Washington, DC: Government Printing Office, May 1992), 5.

38 Harley and Chittaporn, 9.

39 National Technical Information Service, *Indoor Air Quality Environmental Information Handbook: Radon* (Washington, DC: U.S. Dept. of Energy, Jan. 1986), ch. 6.

40 Ibid., ch. 4, p. 5.

41 Presentation by Harvey Keaton at seminar on "Radon: Its Impact on You and Your Municipality," sponsored by the New Jersey Department of Environmental Protection and the American Association of Radon Scientists and Technologists, Atlantic City, NJ, Nov. 19, 1986.

42 National Technical Information Service, ch. 6, p. 24.

43 Brenner, 142.

44 Seminar on "Radon: Its Impact on You and Your Municipality."

45 Kenneth D. Wiggers, "Cold Climate Radon Mitigation—Opportunities and Perils," presented at the 85th annual meeting of the Air and Waste Management Association, Kansas City, MO, June 21–26, 1992.

46 David Bodansky, "Overview of the Indoor Radon Problem," in *Indoor Radon and Its Hazards*, eds., David Bodansky, Maurice A. Robkin, and David R. Stadler (Seattle: University of Washington Press, 1987), 12.

47 Ibid., 13.

48 David Bodansky, Kenneth L. Jackson, and Josph P. Geraci, "Calculated Lung Cancer Mortality Due to Radon," in ibid., 120.

49 Personal communication; Leonard A. Cole, "Radon Scare—Where's the Proof?" *New York Times*, 6 Oct. 1988, A-31.

50 Bernard L. Cohen, "Expected Indoor Radon Levels in Counties with Very High and Very Low Lung Cancer Rates," *Health Physics*, Vol. 57, No. 6 (Dec. 1989); Janet Raloff, "Radon: Is a Little Good for You?" *Science News*, Vol. 134, No. 16 (Oct. 15, 1988), 254; similarly see John S. Neuberger, Floyd J. Frost, and Kenneth B. Gerald, "Residential Radon Exposure and Lung Cancer: Evidence of an Inverse Association in Washington State," *Journal of Environmental Health*, Vol. 55, No. 3 (Nov.–Dec. 1992), 23–25.

51 Susan M. Conrath, "Study Design as a Determinant of Radon Epidemiological Study Validity," Newsletter, Health Physics Society, Vol. 18, No. 7 (July 1990), 1–3.

52 High Background Radiation Research Group, China, "Health Survey in High Background Radiation Areas in China," *Science*, Vol. 209, No. 4459 (Aug. 22, 1980), 877–80. See Werner Hofmann, Robet Katz, and Zhang Chunxiang, "Lung Cancer Risk at Low Doses of Alpha Particles," *Health Physics*, Vol. 51, No. 4 (Oct. 1986), 457–68.

53 William J. Blot, Zhao-Yi Xu, John D. Boice, Jr., Dong-Zhe Zhao, Betty Jane Stone, Jie Sun, Li-Bing Jing, Joseph F. Fraumeni, Jr., "Indoor Radon and Lung Cancer in China," *Journal of the National Cancer Institute*, Vol. 82, No. 12 (June 20, 1990), 1025–30.

54 Conrath, 4.

55 U.S. Department of Energy, Office of Health and Environmental Research, and the Commission of European Communities, Radiation Protection Programme, *International Workshop on Residential Radon Epidemiology*, (Washington, DC: U.S. Dept. of Energy, July 1989); U.S. Department of Energy and the Commission of European Communities, *Second International Workshop on Residential Radon Studies*, July 1991 (Washington, DC: U.S. Dept. of Energy, in press).

56 *A Case-Control Study of Radon and Lung Cancer among New Jersey Women*, Technical Report—Phase I (New Jersey State Dept. of Health, Aug. 1989), iv.

57 Blot et al., 1025.

58 E. Ruosteenoja, *Indoor Radon and Risk of Lung Cancer: An Epidemiological Study in Finland*, Finnish Center for Radiation and Nuclear Safety (Helsinki: Finnish Government Printing Center, 1991), 97.

59 Goran Pershagen et al., "Residential Radon Exposure and Lung Cancer in Swedish Women," *Health Physics*, Vol. 63, No. 2 (Aug. 1992), 179–86.

60 Interview, July 7, 1992.

61 Jay H. Lubin, Jonathan M. Samet, and Clarice Weinberg, "Design Issues in Epidemiologic Studies of Indoor Exposure to Radon and Risk of Lung Cancer," *Health Physics*, Vol. 59, No. 6 (Dec. 1990), 807–17.

62 Ibid., 814–15.

63 Interview, March 31, 1992.

64 *Membership Handbook for the 1991-1992 Term of the Health Physics Society* (McLean, VA: Health Physics Society, Dec. 1991), 5.

65 Daniel J. Strom, "If Indoor Radon Isn't a Problem, Then There Are No Problems," Newsletter, Health Physics Society, Vol. 19, No. 1 (Jan. 1991), 11.

66 JoAnne D. Martin, "Ignoring Radon Is an Outrage," ibid., 20.

67 William A. Mills, HPS Ad Hoc Working Group for Radon, "1989 Draft Statement of the Society's Ad Hoc Working Group on Radon," ibid., 34.

68 Health Physics Society Position Statement: "Perspectives and Recommendations on Indoor Radon, October 1990," ibid., 3; Genevieve S. Roessler and Ronald L. Kathren, "Health Physics Society Position Papers: How Are They Generated?" ibid., 1.

69 Genevieve S. Roessler, "Responses to the HPS Radon Survey," Newsletter, Health Physics Society, Vol. 19, No. 8 (Aug. 1991), 6.

70 Interview, March 27, 1992.

CHAPTER 3

Leapers
and
Loopers

ATTITUDES TOWARD A PUBLIC POLICY DEALING WITH risk may manifest in four ways. The first is belief that the policy is insufficiently protective and should be made more restrictive; second, that the policy is correct and needs no substantial change; third, that it is overly restrictive and should be relaxed; finally, that no policy at all is warranted.

For many risk issues there is at least theoretical room for all four categories. An exception would be when government policy itself is positioned at one of the two extremes. The Delaney provision in federal law, for example, holds that a food additive that causes cancer in any animal may not be used for humans.[1] In this case the first possible category is moot—the allowable risk is zero, and no policy could be more stringent. Moreover, the categories have little practical meaning when a policy is deemed appropriate by virtually everyone. The exclusion from the American market of certain drugs and chemicals, like thalidomide and DDT, is questioned by virtually no one.

In the matter of indoor radon policy, the four categories are cleanly represented and offer a convenient analytic tool. As a descriptive shorthand, people in the categories may be termed leapers, loopers, lopers, and loppers. This chapter presents profiles of individuals in the first two categories; the next chapter focuses on individuals in the last two.

Leapers

People in the first group are labeled leapers because they would "leap" past existing policy in favor of one that is more stringent and aggressive. Their advocacy commonly translates into several proposals:

alter the threshold at which protective action should take place (more restrictive standards, guidelines, or action levels); include more people or locations in categories that require protection; encourage quicker responses by government authorities who are charged with protecting the public; establish more aggressive programs to educate people about potential dangers and how to minimize them; establish or expand penalties for breaches of mandated behavior. The two leapers sketched here with respect to radon policy are Gary Lyman and Gloria Rains.

Gary H. Lyman

Spokespersons for several groups have expressed the need for an expanded and more active radon policy. They commonly represent companies with financial stakes in the matter, preeminently those that perform radon measurements and mitigation. Environmental groups have also spoken in favor of a more vigorous radon policy, as described in Chapter 7. In searching the roster of leapers, however, few scientists working in the field can be found. One is Gary Lyman of the University of South Florida, who holds joint appointments as professor of medicine and professor of epidemiology and biostatistics.

Now in his mid-forties, Lyman has been on the faculty of the university for some 16 years. He became interested in the radon issue in 1984 as accumulations of phospho-gypsum tailings continued to be piled in central Florida. The tailings are a byproduct of phosphate ore mining and fertilizer production and have higher concentrations of uranium than the native ore. "The stacks are health risks," he said, "and they tend to be put in poor socioeconomic areas. I got involved in health risk monitoring, and I thought there was enough data to justify a more aggressive health protection approach on the part of the EPA."[2]

Discontent with EPA's approach to the phospho-gypsum piles led Lyman to examine the radon issue in general. While applauding EPA for recognizing radon as a potential health threat, he criticizes the agency's "decision to base its recommendations for remedial action on a balance of cost and risk." This, he says, accepts a risk "well in excess of risks we have been willing to accept and regulate in other areas and for other carcinogens."

Lyman has published articles on the connection between cancer and radiation and has testified before government agencies on radon and health.[3] He estimates that lung cancer attributable to radon costs the country at least $1 billion in annual health care expenses "and considerably more in terms of lost productivity."[4] He thus considers the EPA's action level of 4 picocuries "indefensible based on either the risks or costs involved."

What would be an appropriate action level? Something close to natural background, according to Lyman.

I accept the EPA's premise that there has to be some balancing of cost and benefit. What I do not accept is the magnitude of the risk that they have been willing to accept—for what seems to be a pretty modest cost. They have ignored the cost of what happens to these people once they've developed lung cancer. I think where my cost-effective threshold fell when I did my calculations was about 1.7 picocuries per liter of air—something in that range. That would be where what you are spending for remediation is what you are saving in health care costs, around 1.7.

Not only does Lyman want to see the action level lowered at least to 1.7 picocuries, he supports "bringing indoor levels down to outdoor ambient levels as a target." This may not be financially feasible in some homes, he acknowledges, but "where readily achievable, particularly in new home construction, we should be striving for ambient levels."

Lyman has urged other more aggressive radon policies as well. He has called for tax incentives to enhance radon education programs and testing and remediation efforts. He supports mandatory testing of homes at the time of sale or transfer. He believes schools throughout the country should be tested and that periodic testing be required in "affected areas." Finally, he would place "restrictions on the rapid residential development of more severely affected areas in high growth regions until effective construction techniques have been fully tested."[5]

Clearly, Lyman's proposed restrictions would crimp current real estate and building practices. His proposals are all predicated on his cost-effectiveness assumptions, which in turn are based on the quandary that underlies the theme of this book—the true relationship of indoor radon to lung cancer. Suppose the ongoing epidemiological studies fail to show a correlation, I asked Lyman as I did other scientists and policy makers. "I am not looking for very conclusive data from current national studies," he answered. "The power of those studies to demonstrate risk is very low. They will require hundreds of thousands of observations over decades." Meanwhile, Lyman assumes the worst.

The data from the uranium miner studies are unequivocal, and you can basically plot a linear dose-response to lower levels of radon exposure. Unless one believes in a threshold effect where the risk drops off dramatically—which has never been demonstrated—then one has to accept the reality that there is a risk in the average residential structure.

When asked why he thinks some scientists are skeptical about the claims of danger from residential radon, Lyman answers with description rather than explanation.

I read *Science* magazine, and their editorial position on most health risks, including radon, is "there is nothing there." They had an editorial policy since [former editor] Philip Abelson took over that environmental risks have been blown out of proportion. I view that as a very pro-industry stance. I don't think that is Abelson's motivation; it's just that he is conservative or whatever you want to call it.

Lyman's attribution of ideology rather than scientific assessment to account for other scientists' positions is not unique. As is evident from other sketches in this and the next chapter, scientists often ascribe nonscientific motives to people with whom they disagree. This is as true for leapers as it is for loppers.

Gloria C. Rains

While Lyman is coolly critical of EPA policies, Gloria Rains is fervent. In view of the lung cancer risk that ostensibly remains at the 4-picocurie level, the agency's reluctance to lower the action level is "unconscionable," she says.[6]

Despite lack of formal training in science or technology, Rains has nevertheless become conversant about a range of environmental issues. Since 1976 she has chaired Manasota-88, a public interest group concerned with environmental health issues. The Florida-based organization was one of several groups around the country established in 1968 by the U.S. Public Health Service to help define community health problems. After its affiliation with the government ended in the early 1970s, Manasota-88 continued as an independent organization. Although its 2,700 members include out-of-staters, most are residents of Florida.

Manasota-88's environmental interests include land use, waste, and energy, but radon has held a particularly prominent place on its agenda. The organization first addressed the radon question in the mid-1970s, a decade before it became a subject of national interest. The issue was tied to the recognition that phospho-gypsum tailings were emitting the gas. Rains recalls her organization's initial concerns.

We became acutely aware of the radon problems in the 1970s because of the phosphate industry. Since then, thousands of houses have been built on these mined out areas even though the government knew they were hazardous. We tried to get them to restrict housing in these areas, but the building went on anyway.[7]

A conversation with Gloria Rains underscores her deep sense of caring for the public good. Now in her mid-sixties, Rains's long-time advocacy on the radon issue has been recognized statewide. The governor appointed her to the Florida Radon Peer Review Committee, a five-member body established by the state legislature in 1986 to recommend safety standards to the state. Arguing for stricter controls, she was a minority voice on the committee. The *St. Petersburg Times* described her as "the only consumer advocate" on the committee. She was consistently outvoted by other members who represented the real estate and phosphate industries.[8]

Rains has urged Florida legislation to make the indoor action level 2 picocuries per liter and to require radon testing in all buildings before they are sold.[9] Her frequent testimony before federal regulatory agencies and congressional committees carries similar themes, often critical of the EPA. In a 1989 presentation to EPA's Science Advisory Board, for example, she called for more stringent standards and characterized considerations to the contrary as "scientifically irresponsible."[10] Her comments in 1991 to the EPA in consideration of its proposed rule on drinking water included this passage:

> EPA recommends that unreasonable risks to health be set at the top of EPA's risk range that is "generally considered acceptable," a 1-in-10,000 lifetime risk.
>
> Considered acceptable by whom? This 1-in-10,000 lifetime cancer risk is not a longstanding nor carefully considered EPA policy. Rather its adoption was the result of a callous political decision. The relatively recent change in EPA policy is evident when considering the cancer risks considered acceptable for pesticide exposure are 1 in 1 million.[11]

In testimony before the House Subcommittee on Health and the Environment in 1987, Rains offered policy proposals that went far beyond existing regulations. They went further than any official was prepared to endorse. She urged, for example:

> That Congress direct EPA to develop *standards* [original emphasis] limiting radon exposure to natural background levels...
>
> The provision of federal and state tax relief to enable home owners to correct radon problems in existing buildings.
>
> A publicly funded radon testing service to be offered free or at cost...
>
> A requirement that adequate radon monitoring be undertaken in all schools and homes prior to occupancy.
>
> [Where high radon levels cannot be mitigated], residential, school and other high occupancy structures may need to be banned.[12]

Drastic as these proposals seem, perhaps Rains's most provocative suggestion concerned the Department of Energy. While urging that the EPA should spend far more money on radon, Rains and her organization urged "removal from lead roles in the radon program of vested interest agencies such as the Department of Energy."[13]

Rains believes the EPA's inadequate policies are a reflection of political misdirection. "The national administration," she concludes, "is sacrificing health for business interests."

Loopers

Loopers, people in the second group in the continuum, reflect the establishment view. They sit uncritically inside the official "loop." The prototypical loopers are the authors or administrators of the existing policy. Their views are preponderant in the regulatory agency where the policy was designed and where responsibility lies for its promotion and enforcement. The term loopers also applies to like-minded people elsewhere. They may be found in the executive and legislative branches as well as outside the government—individuals or representatives of groups that support the official policy. Three loopers are sketched here: Richard Guimond, William Hendee, and Stephen Page.

Richard J. Guimond

No individual has been more influential in shaping the nation's radon policy than Richard J. Guimond. In 1986, he became director of the newly created radon division in EPA's Office of Radiation Programs. He served in that capacity for only 2 years, but they were the years that the radon issue broke into national prominence and that EPA's pre-eminent role in the matter was mandated by federal legislation. Moreover, Guimond's subsequent assignment as manager of all EPA's radiation protection programs enabled him to continue influencing the direction of radon policy.

Guimond holds masters degrees in nuclear engineering and environmental health and is a commissioned officer in the U.S. Public Health Service. After joining the agency in the mid-1970s, he held various positions concerning toxic substances and radiation management before becoming head of the radon division and, in 1988, manager of all radiation protection activities. The following year, while continuing work at the EPA, he was promoted to assistant surgeon general in the Public Health Service with the rank of rear admiral. In 1992 he became assistant administrator of the EPA in charge of Superfund management.

Guimond's radon activities have involved frequent interaction with congressional representatives and their staffs. He developed close relationships in particular with Senator Frank Lautenberg and Congressman Henry Waxman, the chairmen of subcommittees that held several hearings on radon policy since 1986. From that vantage point, he was central to the development of legislation and the consequent responsibilities placed under his agency's authority.

People who have worked with Guimond credit him with having unusual selling skills. He is admired by some for his "great entrepreneurial instincts," as an EPA staff member put it. "He took the radon issue in the mid-1980s and ran with it." This becomes evident in personal conversation as well as in public presentations. Guimond is thoroughly versed in his subject and makes his case in a pleasant and folksy manner.

Although succeeded as director of the radon division in 1988 by Margo Oge, who in 1991 was followed by Stephen Page, Guimond retained his stature as an EPA pointman on radon policy. This was evident at a workshop on "Radon Today: The Science and the Politics," sponsored by the Department of Energy in April 1991. Several EPA radon officials, though not Oge or Page, were in attendance. Guimond offered a formal presentation on behalf of the agency and took the lead throughout the day in response to challenges and questions about EPA's position.

In his early forties, Guimond is personable and fastidiously neat in appearance. His white Public Health Service uniform adorned with rear admiral insignias stood out among the civilian attire worn by other attendees at the workshop. When I asked an EPA radon specialist whether Guimond wore his military uniform frequently, the answer was "Only for big meetings." The audience of scientists and science writers seemed unfazed, as was apparent from their challenging observations about EPA's policies.

Guimond tries to simplify the radon policy matter. "We are down to two questions," he says. "First, is radon a problem as bad as some people say it is? Second, are the nation's policies appropriate?" His answer to both questions is an unhesitant yes. Concerning the first, he says:

> There is a lot of consensus in the scientific community that radon is bad. How bad? Whether it's 16 thousand or 12 thousand [estimated deaths per year from indoor radon exposure] it is a large number. I think most people agree that is a large number compared to a lot of other things we deal with.[14]

His answer to his second question is based on EPA's advice to have a short-term test, and he implicitly responds to criticism that such tests could be misleading.

> If you take an initial short-term measurement in the basement you are unlikely to underestimate the radon problem. With a screening measurement, 80 percent of the people will find they don't have to do any more.... The 20 percent that exceed the action level will have

to gather more data and see if they have elevated levels for an extended period of time.

The title of Guimond's talk at the workshop was "Is Our Nation Doing Enough About Radon?" and his thematic answer was no. He said that animal studies and human data confirm that radon risk estimates carry a "much greater degree of certainty than most environmental pollutants." He listed several presumptions that underlay national policy as set forth by the EPA.

(1) Radon is the second leading cause of lung cancer in the United States, and EPA's "central estimate" is about 16,000 annual lung cancer deaths.

(2) To estimate risks, EPA uses annual average radon levels, not short-term measurements taken in the basement.

(3) EPA estimates that up to 8 million homes have annual levels above 4 picocuries per liter of air, which means about 10 percent of all houses in the nation.

(4) Radiation exposure at 4 picocuries per liter presents a lung cancer risk of about 4 in 100 to smokers and about 4 in 1,000 to people who never smoked.

These contentions, among others, were challenged in the course of discussion, most vigorously by Rosalyn Yalow, whose views are described in the following chapter. The flavor of their differences may be discerned from an exchange about the importance of smoking to the 16,000 annual lung cancer deaths that EPA presumes are a consequence of radon exposure.

Guimond: Some of that population, 15 percent or so, are due to radon alone. The remainder of that group are due to the interaction of radon and smoking. If you eliminate radon altogether, since radon was associated with all 16,000, that would be 16,000 less cases. OK? If you eliminated smoking from that group, the ones associated with radon *and* smoking would go away, but you would be left with 15 percent of 16,000. If you can eliminate one or the other you make a significant dent in the numbers. The only way you'd get rid of it all is to eliminate the radon.

Yalow's response: In the absence of smoking, the age adjusted lung cancer deaths in our country would be between 2 and 3 per 100,000, as opposed to about 55 per 100,000 now. [Her estimate of 2 or 3 per 100,000 is based on U.S. mortality figures before much of the population took up smoking.] In the absence of smoking you would have perhaps 7,000 lung cancer deaths a year. Clearly, only a small fraction, if any, can be associated with radon, because only a small fraction of homes contain a significant amount of radon. If we could eliminate smoking, this very expensive [EPA] radon campaign might save one or two hundred lives a year.

Guimond challenged her assumptions and wondered "how you generated those numbers." With rising emotion, Yalow seemed to take umbrage and amplified her position, which is recounted in the following chapter.

In exchanges with others, Guimond held that scientists in general support an aggressive radon policy. He noted that in recent congressional hearings "we heard various scientific views—but the bottom line was that all the credible scientists were still saying this is a significant health risk." From this he concluded that EPA has not "overreacted" and that its policies are "on track."

Guimond, of course, recognizes that long-term testing gives more accurate results than a 2-day measure. But he defends the short-term approach "to rule out the 80 percent of homes that do not have a problem." With the remaining people "you kind of have a follow up to get a better idea of whether they truly have a problem, and then take action." At the same time, Guimond points out that people do not willingly have long-term tests: "We have studies that show less than 9 percent of the public will put a tester in their home for a year." He concludes, therefore, that despite the weaknesses of short-term testing, it is better than none at all.

No issue was more sensitive than the EPA/Advertising Council campaign, and Guimond acknowledged that it had created difficulties: "Without doubt the unsettlement of the original Ad Council campaign troubled a lot of scientists." But he did not consider the approach a mistake, despite the charges that the campaign was unduly alarmist. "It was in fact effective in getting the message across to people that radon is a concern, and you want to test your house."

When Guimond made this statement at the Science Writers Workshop, physicist Anthony Nero responded that "the Ad Council Campaign isn't anything but an open scandal. If you don't understand this yet, you are still in trouble." In what had come close to becoming a "so's-your-old-man" exchange, Guimond rejected Nero's characterization: "I know that [EPA Administrator] Bill Reilly himself, and other senior people, have looked at it and they do not believe it is scandalous."

The underlying assumption of national policy lies in an Ad Council chart showing a linear relationship between lung cancer and radon at very low levels. Guimond defended the chart as having been based on findings of the National Academy of Sciences and the EPA's Science Advisory Board. "I don't understand," said a member of the audience with reference to the chart, "because even at 2 picocuries you are saying that one person out of every hundred is going to die of radon." Guimond's response: "Right. Radon is a significant risk at virtually any exposure level."

During Guimond's formal presentation he showed a slide titled "Distribution of Radon Risks in Homes." The presumed number of houses where radon levels are above 4 picocuries account for only 33 percent of the lung cancer risk from indoor radon, according to his slide. Thus, if all homes above 4 picocuries were remediated, the risk would be eliminated only for one-third of EPA's estimated 16,000 radon-cancer victims. The remainder would die from exposure to concentrations lower than 4 picocuries. Even if fully implemented, EPA's advisory would by its own presumption leave 11,000 people to die from indoor radon exposure. It is this presumption that inspired Guimond and the EPA to support the national long-term goal of reducing indoor radon concentrations to outdoor levels.

William R. Hendee

Several nongovernmental health organizations have embraced the EPA approach to radon, including the American Medical Association (AMA). The association's position was written by William R. Hendee, the AMA's vice president for science and technology from 1985 to 1991. Hendee, a radiation physicist, held hospital teaching and administrative positions before coming to the AMA in 1985. He left the association in 1991 to become associate dean at the Medical College of Wisconsin. During his time at the AMA he coauthored several articles on radon. Two appeared as reports of the AMA's Council on Scientific Affairs in 1987 and 1991, and both were adopted by the House of Delegates as the AMA's official position.[15]

The reports endorse the 4-picocurie action level and other EPA advisories. While acknowledging the need for more information about the effects of indoor radon, the first report recommends that physicians "should assume a leadership role" in educating the public on the subject. Similarly, the second report calls on physicians to "advise patients and the public" about radon and its health effects. They should help their patients "make intelligent decisions and take responsible action on this issue."[16] These decisions and actions, the AMA directive makes clear, are in line with the EPA's advisories to the public.

Hendee believes that physicians are uniquely placed to offer such advice and potentially are the prime source of information about radon for the public. "Patients ask their doctors about radon," Hendee says in an interview, "because who else do you ask about something possibly impacting on your health."[17]

Under Hendee's direction, the AMA sponsored regional forums to educate physicians about radon. Between 1988 and 1991, 20 conferences were held around the country, each drawing between 50 and 150 attendees. The funding for all the conferences as well as for radon information kits and brochures with the AMA imprimatur on them was provided by the EPA. Grants from the agency from 1988 through 1991 exceeded $250,000. The only cost to the medical association was in the form of staff time.

Hendee said that he and the AMA hoped the newly informed doctors would influence their patients' attitude about radon. If they had any effect, however, it has been scarcely discernible. Hendee shares with EPA officials a sense of frustration that only 5 percent of the public have had their homes tested for radon.

Now in his early fifties, Hendee is a prolific author in areas of radiation physics. He speaks slowly and carefully, and seems perplexed that any informed person would question the wisdom of the nation's radon policy. He is impatient with scientists and others who doubt that radon is a "substantial contributor" to the incidence of lung cancer. He has become convinced that it is because

> to argue otherwise means you have to argue against the science. You have to say really stupid things like "I never knew anybody who died of lung cancer caused by radon." Or, "Show me the bodies." These are very unscientific statements unfortunately made by rather scientific people.

Hendee bases his belief on current radiation models.

> It is true that not all the science and not all the data are in on low level radiation. The evidence that is in, however, suggests that radon really is a contributing factor to the incidence of lung cancer. If you accept the models and the data, and then try to estimate the risk associated with radon exposure, you come to the conclusion as almost every advisory council has—the National Academy of Sciences, the National Council on Radiation Protection, all these. You conclude that radon is a contributing factor to lung cancer. It's not as large a contributor as tobacco, but it is a substantial contributor.

Hendee believes that the EPA's approach to radon has been "judicious." He thinks the agency has been abused by the press "and by some opportunistic individuals who have voiced their opinions through the press, and in the process have represented the EPA as saying things that in fact it did not really say."

What specifically does he mean? He does not mention names, but says that individuals have contended falsely that the EPA urges "quick" measurements "to indicate whether or not you have a radon problem."

In fact, he says, "never once" did the agency propose this. Rather, it suggests short-term screening only "to see if you might have a problem that would then lead to more extensive tests."

A reading of the EPA's 1986 *A Citizen's Guide to Radon* shows Hendee to be technically correct. "The screening measurement only serves to indicate the *potential* [original emphasis] for a radon problem," according to the *Guide*. At the same time, the *Guide* implies that the short-term measurement may itself be taken as the final word, especially if the reading is low: "Depending upon the result of your screening measurement, you may need to have follow-up measurements made to give you a better idea of the average radon level in your home."[18] Thus, according to the *Guide*, a long-term test would be unnecessary "depending upon the result of" the short-term measurement. The EPA's revised 1992 *Guide* says flatly that the average results of two short-term tests "may be used to decide whether to fix your home." The EPA's advisory is not as clear-cut as Hendee suggests.

The closest that Hendee came to criticizing the EPA was in an observation that some of the EPA/Ad Council's advertisements "perhaps in retrospect might not have been done." He recalled one that appeared on television: "a family sitting around, and you hear a zap, and then all you see are the skeletons." The image was then related to radon in their home. "It was sort of scary," he says, "and very controversial."

Nevertheless, he supports the EPA's efforts to overcome public apathy.

> I think one of the ways to get at the radon issue is to lay a burden on families—that as homeowners and as parents they are being irresponsible to their children if they don't ensure that they have a healthy environment. I think we have to be more subtle [than the Ad Council campaign], but maybe that is generally the way to do it. Tell them: "As a parent you're not fulfilling your obligation to your children unless you provide them with a safe home."

In May 1992, 4 months after my first conversation with Hendee, I came across a 1987 news story about the AMA's first report on radon. The story interpreted the report as stressing the need for more research to determine exactly what levels of radon are dangerous and the best way to reduce the danger. The article quoted Hendee as saying:

> There is a serious question in what can you really say about the health risks and how specific you can get.... We can make projections based on other models of radiation at low exposures. But it's not certain that radon follows that model at all.[19]

Hendee's comment seemed more at variance with the EPA's position than was evident from our discussion in January 1992. After reading the article, I contacted him and asked, in light of the quoted passage, if his views had changed since 1987. He said they had not. "It is still not clear in my mind that the model is correct. But one would prefer to overestimate than underestimate the risks, and therefore I am comfortable with the model."[20]

Stephen D. Page

In 1991 Stephen D. Page became the third person to head the radon division of the Environmental Protection Agency since its establishment in 1986. His appointment continued a pattern begun with the division's first two directors, Richard Guimond and Margo Oge. All three were young and articulate, and none was a scientist. Guimond's and Oge's academic training was in engineering and Page's in public administration.

Exuding vigor and enthusiasm during an interview soon after his appointment, Page at age 38 recalled his 14 years with the federal government. He began at the Department of Health and Human Services, moved to the EPA in 1984 where he dealt with groundwater protection issues, and since 1986 has been with its radon program. He was present at the beginning of the aggressive policy period and contributed to its formulation.

Initially responsible for starting the agency's training curriculum on radon assessment, including measurement and treatment, Page soon moved to the public information area. That was in 1986, when the first *Citizen's Guide* was being completed. Although he had little to do with writing the *Guide*, he had "a strong hand in helping to develop the Indoor Radon Abatement Act," which he calls "the framework for the program."[21]

Page's appointment came on the heels of particularly harsh criticism of EPA's radon policies. The Indoor Radon Abatement Act of 1988 and the Advertising Council Campaign that began in 1989 and 1990 had become focal points for what critics considered the agency's misguided policies. The new director said he would seek to reduce the antagonism and to stress areas of agreement. His warm demeanor seems well suited to the task. But the substance of his comments showed little sense that the EPA's earlier approaches were mistaken or that impending policies would take a different course.

In an address at the 1991 annual meeting of the Health Physics Society, Page referred to his reading of a recent society newsletter devoted to the issue of radon. "I was surprised to see how much agreement there was on the science," he said. He thought the articles showed "remarkable agreement" that annual lung cancer deaths from radon were between 5,000 and 30,000. As long as the number was within that range, the agency's policy would not fundamentally change because indoor radon "would still rank as the most serious environmental health threat." In fact, the newsletter's articles offered a variety of views, including harsh criticism of EPA's policies to which Page did not allude.[22]

Page told the assembled physicists that EPA had recently revised downward by 20 to 30 percent its estimates of lung dosage in the home compared to the mine. Previously, doses in the mines and in the homes had been calculated to be equivalent. The revised finding was a consequence of a recent National Academy of Sciences study. In terms of EPA policy, however, this would make no difference, because whatever the actual number, it remained in the thousands.

While acknowledging that "there is room for argument" about the precise level of health risk from indoor radon, Page is certain that the EPA's assumptions are fundamentally correct. Doubts about these assumptions remind him of earlier skepticism about the relationship between smoking and health.

> Back in the 1950s when people started saying that rising lung cancer rates were being caused by smoking, a lot of people said "Don't do anything until we do epidemiological studies." Nevertheless, a lot of scientists were saying the evidence seemed conclusive, and we should do something. Now, I am aware that the tobacco institute and others even today say the correlation between lung cancer and smoking has not been definitely proved. But the insights of the scientific community 30 years ago, and of the Surgeon General, have come to bear the truth.

Page is convinced that radon policy is on a similar track. He may prove to be right. But criticism about the EPA's approach to radon policy seems more widespread among scientists than was true 30 years ago for smoking. Then the vanguard of "scientific" support for the tobacco industry came from scientists employed by tobacco interests. When asked about skepticism on the radon question among scientists who have no pecuniary interest in radon, he responds without hesitation. The issue is influenced by "where you sit and what your responsibilities are." He sees the matter in terms of a "culture clash"

> in the sense that a lot of people, especially health physicists, have made their livings convincing the public that radiation does have beneficial effects. And, by the way I, also believe that x-rays and other radiation have beneficial uses. But these scientists have been

observing the public for years having undue fear about these things even though they as experts know the real risk to be quite low. Now here comes the radon issue, and we are basically telling people that this radiation stuff is really bad. I see a reluctance in the health physics community and the scientific community to say that. You hear it expressed in different ways. For example: "Yes, but by doing this we are going to scare people off from using the beneficial uses of radiation."

Page believes that few scientists dispute the fact that some houses contain radiation levels higher than those from x-ray machines or nuclear power plants. He thinks the critics are fundamentally concerned about communications—how the message is being sent to the public. Many in the scientific community "would like the science laid out in a very rational, informational way, and let the public do with it what they will."

That is the core of the difference that scientists have with the EPA, according to Page. But communications studies have convinced him that the public does not respond to the rational, informational approach. The small number of homes that have been tested for radon stands as confirmation of weak public response. He therefore feels that more emotion in the EPA's message is appropriate even if it "bugs the scientists." How else, according to his reasoning, will the public be prompted to act in its own best interest?

While affirming that EPA's approach has been basically correct, Page acknowledges that "perhaps in looking back we could have sharpened some of the message." But that is the extent of his criticism of the agency's actions. Rather, he sees the issue as complex and therefore subject to being "garbled." He blames the press in part, since reporters tend to "set up a story" by looking for people who would disagree with a given position: "This is the 'pro' side; well what's the 'con' side." The press will seek people who "disagree with the EPA or disagree with the overall policy," Page says, "and then a couple of things start happening." The EPA's message is not clearly delivered and consequently, he believes, even some scientists do not receive the proper facts. But beyond the issue of genuine misunderstandings, he questions the motivation of several in the scientific community.

It is clear to me that there are some people, and this is a minority, who kind of use the situation for their own personal gain. For example, some scientists who say the EPA is using screening measurements in the basement as a basis for mitigation recommendations know that is not true.

Page was unsure of these scientists' motivations, but speculated that by being critical, "they know they can get into the newspapers."

From a scientific standpoint no radon policy has come under more criticism than the Indoor Radon Abatement Act's proclamation of a long-term national goal to reduce indoor levels to outdoor levels. Almost all scientists think such a target prohibitively expensive and perhaps impossible to achieve. Page is defensive about the outdoor-level target and explains the rationale. When the bill was being considered, "the health side of the equation" was being addressed by Congressman Henry Waxman, chairman of the Subcommittee on Health and the Environment of the Committee on Energy and Commerce. According to Page, Waxman initially wanted

> to get this down to zero because it's a carcinogen, and that would be consistent with other carcinogens that EPA regulates. We explained to his staff that measurement at a level of 2 picocuries starts getting real questionable, and if you are talking about ambient levels, even perhaps 1.5, you are really into a gray area.

According to Page, Waxman responded by saying, in effect, "EPA, I hear what you are saying about ambient and zero, but you ought to be telling people that there are risks below 4 picocuries, and not take the incentive away from industry to stop at 4." Page continues:

> So the conversations and the negotiations went back and forth between the higher ups in this agency and on Capitol Hill, and in the waning moments Waxman was finally talked into a national goal.... This thing was on a much different track than anybody ever knew, and I think Rich Guimond [then director of the radon division] has taken an unfair amount of criticism for putting him up to this. I know he didn't put him up to this because I was there.

In the end, Page stands by the EPA's policy of aggressively encouraging every homeowner to test for radon. His justification, presented with earnest conviction, is that "EPA is concerned about the individual citizen." He dismisses a policy alternative that would amount to "fumbling around for 5 or 10 years to find all the homes over 20 picocuries or whatever."

How representative is the loopers' position? While impossible to specify the exact number of individuals in the various categories, there are evidently more loopers than leapers. The nature of being a leaper about any risk policy—a protagonist for more aggressive action—means one is likely to speak out and be visible. Leapers, as individuals and as representatives of groups, are prone toward activism and making their positions known.

Thus, in the matter of radon policy, the standing of leapers may be gauged by the frequency that their position is expressed at forums, at congressional hearings, in statements to the media, and in professional and general publications. An overview of these vehicles for expression leaves little doubt that leapers are far fewer than those in the loop.

Notes

1   Title 21, §348, enacted in 1962, includes the provision that "no [food] additive shall be deemed to be safe if it is found to induce cancer when ingested by man or animal."

2   Unless otherwise noted, quotations in this sketch are from an interview with Gary H. Lyman, July 10, 1991.

3   Gary H. Lyman, Carolyn G. Lyman, and Wallace Johnson, "Association of Leukemia with Radium Groundwater Contamination," *Journal of the American Medical Association*, Vol. 254, No. 5 (Aug. 2, 1985); Heather G. Stockwell et al., "Lung Cancer in Florida," *American Journal of Epidemiology*, Vol. 128, No. 1 (1988).

4   Gary H. Lyman, "Environmental Risks for Lung Cancer" (statement to the National Cancer Advisory Board) (Feb. 11, 1988), 7.

5   Ibid., Hearing, National Cancer Advisory Board, 8–9; House Subcomm. on Health and the Environment of the Comm. on Energy and Commerce, *Hearing on Radon Exposure: Human Health Threat*, Nov. 5, 1987 (Washington, DC: Government Printing Office, 1988), Testimony, 92.

6   Ibid., 83 (statement by Gloria C. Rains).

7   Unless otherwise noted, quotations are from an interview with Gloria C. Rains, Nov. 18, 1991.

8   Editorial, *St. Petersburg Times*, 10 Dec. 1987, 26–A.

9   Ibid.

10   Gloria C. Rains, Presentation to the Radiation Advisory Committee of the Science Advisory Board, Environmental Protection Agency, Apr. 26, 1989.

11   Gloria C. Rains, Comments on Proposed Rule: National Primary Drinking Water Regulations; Radionuclides, 40 C.F.R. pts. 141 and 142 (July 18, 1991).

12   *Hearing on Radon Exposure: Human Health Threat*, 87.

13   Ibid.

14   Unless otherwise noted, quotations and information in this sketch are from Guimond's presentation and conversation at the Science Writers Workshop on "Radon Today: The Science and the Politics," sponsored by the U.S. Department of Energy in Bethesda, MD, Apr. 25–26, 1991; and from an interview, June 11, 1992.

15   Theodore. C. Doege and William Hendee, "Radon in Homes," Council on Scientific Affairs of the American Medical Association, *Journal of the American Medical Association*, Vol. 258, No. 5 (Aug. 7, 1987), 668–72; ibid., "Health Effects of Radon Exposure," *Archives of Internal Medicine*, Vol. 151 (Apr. 1991), 674–77.

16   "Radon in Homes," 672; "Health Effects of Radon Exposure," 677.

17   Unless otherwise noted, quotations and information are from an interview, Jan. 3, 1992.

18   U.S. Environmental Protection Agency, *A Citizen's Guide to Radon: What It Is and What to Do About It* (Washington, DC: U.S. Environmental Protection Agency, Aug. 1986).

19 "AMA Says Facts on Radon still Unclear," *Record* (Hackensack, NJ), 21 Aug. 1987, C-16.

20 Interview, May 4, 1992.

21 Quotations and information are from presentation at the 36th annual meeting of the Health Physics Society, Washington, DC, July 23, 1991; and from an interview, July 25, 1991.

22 Health Physics Society, Newsletter, Vol. 19, No. 1 (Jan. 1991).

CHAPTER 4

Lopers
and
Loppers

AFTER PROPOSING AN ANALYTIC SCHEME TO ASSESS DIF-
fering approaches to radon policy, the previous chapter presented the
views of leapers and loopers. This chapter focuses on the remaining
two categories, lopers and loppers.

Leapers feel that EPA radon policies are not sufficiently stringent.
But the great majority of critics are people who think the agency has
gone too far, too fast. They broadly fit into two groups—those who
believe a more limited national radon policy is warranted and those
who think no policy at all is appropriate.

Lopers

Lopers see existing policy as overly stringent. They think the policy
should be modified because as presently constituted it promotes
unnecessary worry among the public, wastes public and private money,
and diverts resources from indisputable needs. People in this group
exhibit a lower or narrower sense of urgency than those in the loop.
They counsel hesitancy until more information is available to confirm
the suppositions upon which official policy has been structured. Some
favor focused efforts to locate the limited number of homes at very
high risk levels. Others minimize the need for aggressive action of any
sort until more is known about the subject. Collectively, none runs
with official policy but rather they "lope" behind. The three lopers
profiled here are Ernest Létourneau, Susan Rose, and Anthony Nero.

Ernest G. Létourneau

Although a Canadian, Ernest G. Létourneau is a prototypical loper as the term applies to U.S. radon policy. Few U.S. or Canadian regulators have been involved longer in the assessment of risks from radon to the general population. Until the mid-1980s Canada's permissible indoor radon level pertained only to homes in uranium mining areas and was 4 picocuries per liter of air. But in 1988, national and provincial authorities established a standard for houses throughout the country at 20 picocuries per liter. Létourneau thinks the standard prudent and wise. Indeed, he has been a central figure in the development of Canada's radon policy.[1]

Létourneau is in his early fifties and displays a ready smile. His expression turns impish as he suggests that the reason fewer than 5 percent of U.S. homeowners have tested for radon is because "they have more sense than the Environmental Protection Agency."

A physician by training, Létourneau has devoted most of his professional life to the study of radiation effects on humans. In 1968 he joined the radiation protection division of Canada's Department of National Health and Welfare, and since 1978 he has been director of the Radiation Protection Bureau. He began working on the radon issue in 1976, conducted the first countrywide radon survey soon after, and subsequently chaired the committee that developed Canada's federal-provincial radon policy.

Expressing confidence that Canada's policy approach to radon is correct, Létourneau speaks with exuberance. He recites a litany of contradictions about radon, including mistaken beliefs about its concentration and effects. He has found, for example that contrary to popular thinking, insulating a home does not increase radon concentrations. "In fact, we have found that the oldest houses in Canada, the leakiest houses, have the highest radon levels." The New Jersey epidemiological study that was cited by the state's Department of Health as confirming a relationship between residential radon and lung cancers "really shows nothing," according to Létourneau. "It is nonsense to say they found a significant trend, but in the same study say that the relationship between lung cancer and radon levels was not found to be statistically significant."

Létourneau then refers to a Canadian study of 750 lung cancer victims and 750 controls that has been in progress since 1984. The study included an effort to locate every resident of Winnipeg who contracted lung cancer, trace the homes they lived in, and then do long-term radon measurements in those homes. Preliminary findings are expected by 1993. But Létourneau makes a pointed observation concerning what is already known about the experimental group: "Frankly, after finding the 750 experimental cases [all the lung cancer victims in Winnipeg], we found that every one of them was a smoker. Where are the nonsmoker lung cancer victims, the ones who might have contracted cancer just from radon? I don't know. We could not find any."

Létourneau does not dismiss the need for concern about radon, but he thinks the U.S. policy is too aggressive. He is persuaded by animal studies and miner studies that high indoor levels should be mitigated. "Even if our study shows no measurable relationship between lung cancer and radon, Canada will continue to have a radon policy." What will the policy be? The study should confirm one of two possibilities. "Either the risk estimates we use, which are accepted world-wide, have resulted in cancers, or cancers do not appear at that level and therefore this is a ceiling for risk estimates." He expects that the findings of the Canadian study will offer "a better handle on what we can do for public policy."

Revealing the quintessential approach of the loper, Létourneau explains the rationale behind Canada's policy.

> The public policy was established by doing a radon survey of the country and studying lung cancer rates in all the major cities, then trying to relate them while controlling for smoking. We did that, and it did not show an effect. This meant on the political level that radon was not a public health hazard in the same way as typhoid or any other big thing. So our attitude was, let's study the risk and not yet spend a lot of money. We'll wait until we study the risk and then make a decision—should we stay with [the action standard] we have, should we go down, should we go up.

Létourneau notes that seven countries in the world have radon policies, and none has provided funds for measuring radon levels in every house. "That is very important," he says, "because if people truly believed that the risk was high, there is no doubt that governments would offer credit to pay for it out of taxpayers' money." He noted in 1991 that the EPA was considering reducing the U.S. action level to 2 picocuries of radon per liter of air. He says this "would mean a significant number of places in large areas of Canada could not even meet an indoor standard, because natural outdoor levels are higher." Natural outdoor radon levels in Manitoba and Saskatchewan commonly approach 3 picocuries, though lung cancers appear no more frequently in those areas. "How many more surprises we'll have in the radon business remains to be seen," he says.

As Létourneau warms to his subject he distances himself further from the U.S. policy loop.

When you talk about this subject, there are many other attributable causes of lung cancer, and this presents a problem. What do you do with passive smoking? Some of my colleagues in epidemiology say that it is not radon—that these other people are affected by passive smoking. And they say passive smoking has the same risk as environmental radon. Then I'm stuck because I don't have enough dead bodies to go around. I've got too many risk estimates and not enough deaths. Then you have to talk about asbestos and occupational carcinogens, and in addition we know there are some genetic factors.

It gets to be very complicated to deal with. And the mathematical game of course is that of the risk estimate—and it is a game. As scientists we tend to get drunk on beautiful numbers. But those numbers are very weak and fuzzy. They are not hard numbers, and if you play around with the risk estimates you can show all sorts of effects which may or may not truly be there.

Létourneau inferentially touches on his conflicting roles as scientist, regulator, and subject of political pressure. He seems to be a man struggling with the impossible: maintaining a radon policy, but feeling discomfort with the evidence that provides the basis for the policy. He is uneasy about the commonly accepted models that presume effects of low-level radiation exposure based on knowledge about high-level exposure. He comments on the presumed risks derived from the models.

These risk estimates show a one-in-a-thousand risk from breathing normal, pure, uncontaminated air. There is something wrong with that philosophically. Either we are controlling risks which are too small in normal life, or we overestimated the risks for various reasons. But there is something philosophically wrong with living in a [normal] atmosphere which is unsafe.

The patently absurd conclusion, Létourneau says, is that "according to our risk estimates, normal, unpolluted outdoor air is unsafe to breathe."

Susan L. Rose

Létourneau's skepticism may not be reflected in official U.S. policy, but neither are all U.S. policy administrators subscribers to the policy. Susan Rose, manager of radon programs in the Department of Energy (DOE), is outside the official policy loop. It may seem a paradox to describe the highest official dealing with radon in a government agency as a loper. This is especially true because the department was once the principal agency concerned with radon issues. But since the mid-1980s, as a result of the federal radon laws, the EPA has become the locus of radon policy formulation and administration. Differences about approaches concerning radon policy as well as the accretion of power by the EPA have left a residue of tension between the two agencies.

Rose holds a Ph.D. in medical technology and taught biology for 7 years before joining the DOE in 1975. Since 1987 she has managed the department's radon programs as well as its human subjects research policies. She speaks with enthusiasm and is not averse to using spicy language to make a point. She wonders, however, where her views about radon may take her career. She says she was pointedly reminded by an EPA official that Congress and the administration have taken a position on radon, "and we better not get in the way."[2]

Rose was unsettled by the keynote address at an EPA-sponsored symposium urging that uncertainties about radon policy be withheld from the public.[3] Her reaction was "extremely strong," she recalls. While very high radon levels in homes might pose a danger, she thinks that current knowledge does not warrant EPA's 4-picocurie action level. She is most upset about any efforts to close off debate. "I think that public policy is best served when multiple voices are heard."

Emphasizing that no one has the right to claim a single correct answer, Rose is mindful that

> each of us has a vested interest in this. Our message at the Department of Energy is one of uncertainty because that is what we do— we look at scientific questions yet to be answered. The interest of industry is obviously economic and making money. The Environmental Protection Agency has its own interest.

Rose mentions the DOE's long history of radon research, back to the early 1970s. In recent years, however, as the EPA has received more funding for radon research, the DOE has been budgeted for less, down from $13 million in 1990 to $10 million in 1991. Nevertheless, the department maintains a broad program and continues to sponsor radon studies involving geologic formations, instrumentation, entry mechanisms into homes, dosimetry, and epidemiology. Research is performed at national laboratories, academic institutions, and other federal agencies.

Rose oversees ongoing investigations and helps plan and initiate new ones. She brings questions to the policy agenda that, she says, receive little public attention. Neither in the press nor at congressional hearings have her considerations been sought with the frequency of those of EPA spokespersons. She summarizes uncertainties concerning radon that she and the DOE are trying to address.

> The outstanding question, and the place where our program would like to make a contribution, is the linear question at low levels. Is

there a way to find whether there is or is not a linear relationship between radon and lung cancer that goes all the way down to zero?

Another question is where are the hottest homes—those with very high radon concentrations? It would be more expedient if we had a way of identifying them. We are launching into this—to find where the hottest areas may be.

Questions remain about the results of ecological and epidemiological studies. What is the reason they all appear to be showing no relationship at lower levels?

The measurement technology is another area where improvements can be made. If you get a reading of 4.01 picocuries you can be misled. Moreover, is mitigation effectiveness durable? Three years from now will the number be what you paid to have it reduced to?

Another long-term interest at DOE concerns carcinogenicity from radiation. Do you get carcinogenicity from radon without other sources in inhaled air such as from smoking and aerosols from cooking? And what about genetics? In some studies it seems as though you have to have a genetic predisposition to actually get lung cancer.

Susan Rose sums up her philosophy with a plea to trust the good sense of an informed public. "I do not think the consumer has to become a radon expert. But consumers certainly deserve to hear what we do know and what we don't know, and then have the option to make an educated decision about their own homes and their own safety."

Anthony V. Nero, Jr.

The third loper described here, Anthony Nero, is a 50-year-old physicist at Lawrence Berkeley Laboratory of the University of California. He has a long history of research and publishing on the subject of indoor radon. Although he is soft-spoken, his judgment is stern. He thinks the EPA greatly exaggerates the indoor radon problem.

Nero observes that some data issued by the EPA imply that 30 percent of homes in the United States have radon concentrations that exceed 4 picocuries per liter. The data, he contends, are based on screening measurements that "are not indicative of concentrations to which people are actually exposed." The measurements were taken in basements, where radon levels are often higher than elsewhere in a house and where people spend little time. Moreover, short-term testing frequently does not reflect annual averages, he emphasizes. Data he has analyzed indicate that 6 or 7 percent of U.S. homes may have radon levels above 4 picocuries.

Nero figures that a strategy of testing and remediating homes based on a 4-picocurie action level could cost $20 billion, more than twice the EPA's projection. Even this figure understates the costs of a radon program advised in an EPA brochure produced with the Ad Council in 1990, he says. He is especially critical of the brochure's message that "having radon in your home is like exposing your family to hundreds of chest x-rays yearly." No distinction is made between differing levels of radon, he says, nor is the fact mentioned that everyone is exposed to radon in their homes. "Information of this kind with the EPA stamp on it, if fully implemented, could lead to a much bigger program than one for $10 billion or $20 billion. It would be a program on the order of $50 billion or $100 billion."

He is skeptical about the gains from a 4-picocurie program in any case. This would reduce the number of radon-associated lung cancer deaths by about 15 percent, according to Nero. "If you count all the health risks associated with smoking that lead to death, that's equivalent to about a 1 percent reduction in smoking. So the entire EPA program, the $10-to-$20-billion one, if properly executed, would have the effect of a 1 percent reduction in smoking."

Nero then observes that a 2-picocurie action level program, which the EPA/Ad Council brochure implied would be preferable, would reduce deaths by only another 1 percent of the number caused by smoking. But his most pointed cost-benefit critique is reserved for the EPA-endorsed legislation passed by Congress in 1988. Nero says the following:

> The Indoor Radon Abatement Act sets as a long-term goal the reduction of indoor levels to outdoor levels. This would be very costly. Various estimates have come forward in the range of $500 billion to $1 trillion.... And this would give us just another one percent equivalent reduction in smoking.

After pausing for a listener to absorb the impact, Nero smiles and says: "What I think we need is some perspective."

Nero's solution involves targeting houses where radon concentrations exceed 20 picocuries per liter, the occupational standard for workers in radiation-related occupations. About 100,000 houses would be involved, he figures. "It's not an easy job to find them, but we can focus on them more rapidly [by paying] attention to geologic information." Based on existing knowledge about uranium concentrations in the United States, he estimates that "90 percent of these high radon houses might occur in about 10 percent of the country." He advocates "focus programs" in these areas to encourage responsive building codes and to stimulate people to test and mitigate. The cost of such a program, he believes, would be about $500 million.

Nero is concerned that residents in very high radon homes are at risk. He is classified as a loper because his proposed program would focus only on these homes, a far more limited approach than that of current policy. Indeed, he is not only outside the policy loop but is among the most outspoken critics of EPA policies.

An Anonymous Loper

As a postscript to these profiles of lopers, below is a comment from an EPA official who requested anonymity. A closet loper, the official is uncomfortable with the agency's radon policy for reasons that are no less philosophical than pragmatic.

> I guess the basic philosophical question is how intelligent is the American public. Can the American people handle different options—can they handle the fact that risks are uncertain? Or do we have to speak to them on a third grade level, so prescriptively that we almost take on an Orwellian big brother attitude.
>
> I am afraid that the EPA has gone into a public relations blitz which may be appropriate for a group like General Motors that wants to market a product. But I don't think it's an appropriate federal role.
>
> Would we have been any more effective or less effective had we taken a more informational rather than motivational approach? I can't say. But I can tell you that only about 5 percent of the people have gone ahead and measured for radon. If you guess how many among these turned out to be above 4 picocuries and whether or not they mitigated, out of the supposed 20,000 lives lost to radon maybe we saved 50.

This EPA staff member sounded sadly resigned when asked if this view was expressed within the agency. Yes, the official said, but it is a minority view that goes unheeded.

> They always say "Look, we can't please everybody." The majority view is that risk communications are more important than scientific accuracy. The discussions usually end with "O.K., so we heard your comments. We're done."

Loppers

The last category includes people who think no government policy is warranted because the risk from indoor radon is trivial or nonexistent. From their perspective, any officially sanctioned policy is unnecessary, wasteful, and misleading. Proponents of this view would urge that the issue be lopped from the government's agenda and in shorthand may be labeled "loppers." The two loppers sketched here are Rosalyn Yalow and Philip Abelson.

Rosalyn S. Yalow

The most richly credentialed participant in the radon debate is Rosalyn S. Yalow. A health physicist, in 1977 she was awarded the Nobel Prize in Physiology or Medicine for the development of radioimmunoassay, a method of measuring trace substances in the blood and other body fluids. Born in 1921, she has worked at the Bronx Veterans Administration Medical Center since 1947. She has long felt that the public has been made unnecessarily worried about low-level radiation. The radon issue, she believes, exemplifies the latest and perhaps most egregious example of misguided radiation policy.

Yalow's no-nonsense Bronx delivery seems a contrast with the global accomplishments listed in her curriculum vitae. She has received more than 50 awards, in addition to 47 honorary doctorates from universities around the world. With relentless skepticism she confronts anyone who tries to argue that indoor radon is a menace. Presenters at a 1991 forum on the subject were nonplused by her approach. Away from the formal proceedings, she grinned and said that she does not mind being considered a troublemaker. "I enjoy challenging them and showing how ridiculous the EPA's radon program is." To Yalow, science is "just sensible thinking," and she is convinced that knowledgeable scientists understand the folly of the nation's radon policy.[4]

Why then, I ask her, do not more scientists openly oppose the notion of a national radon policy? "Because they would lose their grants if there was no radon program," she responds. "Even some who have criticized the policy in the past have gotten quieter. They are getting their money from EPA and DOE, and if there are no radon programs they won't be getting any more money." Her conjecture may be true for some scientists, though it seems unprovable. But a blanket characterization of the hundreds of investigators doing work on radon seems unfair; I have talked with several who very much seem to believe in the establishment position.

Yalow has expressed her rationale in articles and at scientific meetings. She begins with the premise that radon and its progeny, like other radiation emitters, have always been with us. At the concentrations commonly found in homes, there is no evidence that in their own right they cause harm. Only when present at very high levels, and particularly among smokers, has there been evidence of danger. She concludes that there is "no reproducible evidence of harmful effects associated with increases in background radiation up to 6 times the usual levels."

Yalow cites the study conducted in China that compared the

health of people in high-background radiation locations to those in control areas. Some 73,000 people were investigated who lived in regions where surface soil contained about 2000 picocuries of radium per kilogram. For the control group of 77,000, radium concentrations ranged from 300 to 900 picocuries. Radon levels were not mentioned in the study, but presumably reflected the amount of radium in the soil below.[5] (Soil concentrations of uranium, thorium, and potassium nuclides were also much higher in the experimental than the control areas.)

The areas are only a few miles apart and at comparable altitudes. As discussed in Chapter 2, 90 percent of the high-background inhabitants were from families who had lived in the area for six or more generations. The investigation included examinations for hereditary diseases, malignancies, spontaneous abortions, and more. No statistically significant differences for any disease or disability were evident. During the period under study, 1970 through 1974, one lung cancer death was noted in the high-background area and four in the control area. The authors concluded that either the effects of low-dose radiation are so small as to be undetectable, or they do not exist at all.[6]

Although Yalow highlights the study as important evidence for her argument, she also mentions findings in the United States. She observes that the three states with the highest radon levels in homes are Colorado, North Dakota, and Iowa, where averages range between 3.3 and 3.9 picocuries. The three with the lowest are Delaware, Louisiana, and California, with averages below 1 picocurie. Yet lung cancer deaths in the highest radon states average 41 per 100,000, and in the lowest radon states 66 per 100,000. This, she says, suggests the possibility of an inverse relationship between relatively low radon levels and lung cancer.

This is not the only domestic information the EPA seems to have skirted, Yalow says. "The EPA does not emphasize that its estimates of radon-related lung cancer deaths are based on observations among smoking miners. At the EPA's recommended action level for home radon...it would have taken many lifetimes to accumulate sufficient exposure from radon daughters to have resulted in an increase in lung cancer in non-smoking miners."[7]

That is the crux of Yalow's argument—that smoking is the culprit to be addressed, not radon. She explains at the 1991 Science Writers Workshop:

> In the absence of smoking, the age-adjusted lung cancer deaths in our country would be between 2 and 3 per 100,000 as opposed to about 55 per 100,000 that is now the case. Thus in the absence of smoking you would have about 7,000 lung cancer deaths a year....

Clearly, only a small fraction of these deaths can be associated with radon, if any are associated at all, because only a very small fraction of homes contain a significant amount of radon. If we could eliminate smoking, this very expensive radon campaign that we're talking about might save 100 or 200 lives a year. This is the way you should look at it.

In the middle of her remarks at the workshop, Yalow was interrupted by Richard Guimond, director of the Office of Radiation Programs at the EPA, who wanted to know how she generated those numbers. She responded impatiently, as if the answer should have been obvious. She recounted that in 1930 the lung cancer death rate among women was 2.5 per 100,000, and among men 5 per 100,000, according to American Cancer Society statistics. More members of both sexes began to smoke after this period, leading to today's higher lung cancer rates. She continued:

> In fact if you look at Mormon women in 1970, the rate is still only 4 per 100,000, which is close to the 1930 figure for women—a little higher now perhaps because an occasional Mormon woman was smoking. So we are left with the fact that in the absence of smoking the lung cancer rate is 2 to 3 per 100,000. Then we have to deal with asbestos [and] air pollution, so we have to get rid of some of those numbers.

Yalow is convinced that the money spent on radon should be devoted to reducing smoking or other demonstrable risks such as fires at home. "A campaign to install smoke detectors in residences would be less expensive and save more lives than an expensive campaign to find the very rare home with a high enough radon level to present a significant risk to a nonsmoker."

Philip H. Abelson

While Yalow is unsparing in her criticism of the EPA's radon policy, Philip Abelson injects ridicule. Born in 1913, he has made impressive contributions in a variety of science areas and in 1992 received the Public Welfare Medal, the highest honor given by the National Academy of Sciences. A physicist by training, his earliest work involved identification of products of uranium fission, which led to his codiscovery of the element neptunium. He later turned to research in organic and biochemistry and molecular biology, which included the discovery of amino acids preserved in ancient fossils. Between 1962 and 1984 he was editor of *Science*, the journal of the American Association for the Advancement of Science. Now deputy editor, he writes an occasional column and, in 1990, offered one titled "Uncertainties about the Health Effects of Radon."[8]

Abelson wrote that the cost to homeowners to fulfill the congressional mandate of reducing indoor radon levels to those outdoors would average about $10,000 each. That would mean a national expenditure of $800 billion. His column challenged assumptions behind the government's policy, including the belief that high-dose effects can be extrapolated to predict those at low doses. "That is an assumption that has never been proved." He concludes with the thought that EPA might better focus on the "rare circumstances" where levels of radon are elevated.[9]

Subsequently, however, Abelson suggested that any radon policy was unwarranted and wasteful. At the 1991 Science Writer's Workshop, he offered this view in a paper titled "Radon Today: The Role of Flim-Flam in Public Policy." The paper begins with a dismissal of the EPA's presumption that "large exposures in dusty mines" can be extrapolated to "low doses in relatively dust-free living rooms."[10] Much of the miner lung cancer data involved men who worked in mines in the 1940s and 1950s but whose cancers manifested in subsequent decades.

Uranium miners in the Colorado Plateau during this period labored under inordinately primitive and unhealthy conditions, Abelson reported. He had visited several mines at the time and received additional information from U.S. mining and health officials then in service. After World War II many mines were started by amateur prospectors seeking the newly valued uranium ore. "In many cases the openings were scaled to a size more comfortable for dogs than humans," he says. In these cramped conditions, where most of the miners smoked, everyone present "shared abundantly in the smoke" as well as exposure to nitrogen oxides and mineral dusts that contained uranium and its decay products. "The miners—both smokers and nonsmokers," he says dryly, "were exposed to hazards not present in your home."

He concludes that the circumstances of the miners—the lack of reliable air quality measurements, the work environment, the paucity of nonsmokers—further weaken a case for extrapolation.

> Most of the citizens of the United States are nonsmokers. When EPA depends on data from the miners to justify frightening nonsmokers it is engaged in a highly dubious enterprise. In summary, extrapolation of questionable data from mines is of doubtful value as a foundation for national policy involving a possible trillion dollars or more.

Abelson then turns to the lack of epidemiological evidence, noting that the EPA has estimated as an upper limit that among 140,000 annual lung cancer deaths, 43,200 might be due to radon.[11] "Such a large number—whether 43,200, 20,000, or 16,000—should be glaringly evident from even a casual epidemiological survey." He refers to studies, like the ones noted by Yalow, that show lower than average lung cancer rates in states with the highest radon levels. The statistics "seem to demonstrate that, if anything, moderate levels of radon are beneficial to the public health."

Abelson moves this theme from apparent tongue-in-cheek to consideration of its actual validity. He recalls that the linear hypothesis suggesting harm from small doses of radiation was developed 50 years earlier in the absence of solid data. Current data indicate otherwise. While ionizing radiation can injure chromosomes, "it is now known that repair mechanisms exist." An adaptive response may exist that can explain why "low level radiations make the cells less susceptible to subsequent high doses of radiation."

He notes that several elements essential to life are carcinogenic at high levels. Excessive ingestion of table salt, for example, can cause stomach cancer. "If EPA were consistent in its regulatory program," he says," the known occurrence of salt-induced stomach cancer should cause a banning of table salt for human use."

Abelson argues that the EPA has no data to assume that human responses to radiation are different, and he rejects the agency's radon program as an effort to "brainwash the public." The EPA appears to think its goal "justifies using inaccurate data and inflicting psychological trauma." In his remarks, Abelson describes the agency's approach as "shrill," "inexcusable," and "designed to raise anxiety rather than present facts." He excoriates the EPA for trying to scare mothers into action by alleging that children are three times more susceptible than adults to harm from radon. "The model for respiratory cancer," he notes, with reference to the National Academy of Sciences' BEIR V report on exposure to low levels of ionizing radiation, "does not depend upon age at exposure."[12]

Abelson shows further examples of what he considers EPA's excessive zeal by mentioning a briefing document, "Radon Media Campaign," concerning the agency's relationship with the Advertising Council. Recommendations included: "Eliminate unnecessary information" and "[u]se strong and unsettling messages." "In other words," he says, "don't inform them, scare them."

The EPA, according to Abelson, has established a pattern over the years: As a risk is identified, the agency gives credence to "sloppy data if they indicate a greater risk. Experiments that later show that no risk exists are disregarded." This pattern, he suggests, is now forming the agency's approach to radon.

Abelson's final and most acerbic condemnation came in an impromptu remark after his formal delivery was completed. "The EPA is in the process of destroying the credibility of the federal government."

Impressions

Unswervingly dedicated to their convictions, loppers are far fewer in number than lopers. Thus, the four categories are not evenly represented. The continuum bulges at the center among loopers and lopers—supporters of the establishment position and advocates of a more limited or focused approach. Moreover, more people seem to support the EPA policy, even if tepidly, than criticize it.

While the four categories offer clear distinctions on a continuum of attitudes, any manner of applying labels to people may be subject to exceptions. Not all people in a single category think exactly alike. Also, individuals may show some characteristics that fit a particular category, yet other characteristics that apply to another. Thus, a person might generally support EPA programs, but prefer a slightly more aggressive approach. This individual straddles two categories, leaper and looper, and defies an effort to label precisely. Nevertheless, people attentive to the radon issue tend to fit into one of the four categories, which allows for a more systematic understanding of the range of views.

In assessing the sketches in this and the previous chapter, the overwhelming impression is the strength of conviction held by the protagonists. Whether government official, scientist, or community activist, they implicitly, and sometimes explicitly, present their cases as if none other could be valid. In some respects this is understandable. It reflects a human propensity to validate a position that one has adopted. This is particularly true for government officials who have a responsibility to act. They cannot, after all, be urging the citizenry to spend billions and at the same time appear to doubt that the policy they are advocating is correct. But this observation is no help in the quest to determine the best policy, the theme to which subsequent chapters return.

It is no coincidence that the sketches of leapers and loopers included fewer scientists than those of lopers and loppers. The selections were an attempt to represent fairly the backgrounds of people who have spoken prominently for each category. The major exception to this was the exclusion of individuals who had an obvious financial interest in a position. Many representatives of radon testing and mitigating companies are leapers. But their convictions may justifiably be suspected insofar as they promote their economic interest. This applies similarly to representatives of real estate interests who are amply found among lopers and loppers.

In seeking to profile people who apparently would not gain financially from the policies they espouse, a disproportionate number of scientists exists among lopers and loppers. This does not signify a dearth of scientists among loopers. Many scientists support the EPA's position, including members of the Radiation Advisory Committee of the agency's Science Advisory Board. But the most vociferous and publicly visible loopers are those who make policy, and within the agency that means nonscientists.

Is that appropriate? Should not scientists, who know most about the subject, have the final word on public policy? Certainly not. In the first place, not all scientists agree on policy, as is evident from this and the previous chapter. Even if they did, scientists should no sooner be able to dictate policies about radon than they should about the adoption of nuclear power, genetic engineering, or agriculture.

Public policies should not be subject to the decree of scientists, regulators, or any particular group. They should derive from the marketplace of ideas where competing interests are heard in open forum. The process is part of democratic politics.

The contribution of congressional committees can be instrumental to the process. When operating at their best, committee hearings are valuable vehicles for spokespersons from a variety of interests to offer views about policies—to advocate and educate. The hearings can help the public and their elected representatives learn about issues of importance to the country. It is this forum that is explored next in examining the roles of federal regulatory and elected officials.

Notes

1   Unless otherwise noted, information and quotations by the people profiled in this chapter are from presentations and conversations at the Science Writers Workshop on "Radon Today: The Science and the Politics," sponsored by the U.S. Department of Energy in Bethesda, MD, Apr. 25–26, 1991. Additional information about Ernest Létourneau is from an interview, May 6, 1992.

2   Interview, Feb. 11, 1992.

3   See John R. Garrison, Managing Director, American Lung Association, Keynote Address, Environmental Protection Agency, *The 1991 International Radon Symposium on Radon and Radon Reduction Technology*, Philadelphia, PA, Apr. 1991, discussed in Chapter 1.

4   Interview, Apr. 26, 1991.

5   High Background Radiation Research Group, China, "Health Survey in High Background Radiation Areas in China," *Science*, Vol. 209, No. 4459 (Aug. 1980), 877–80.

6   Ibid., 879–80.

7   Rosalyn S. Yalow, "Concerns with Low Level Ionizing Radiation: Rational or Phobic," *Journal of Nuclear Medicine*, Vol. 31, No. 7 (July 1990), 17A.

8   Philip H. Abelson, "Uncertainties about the Health Effects of Radon," *Science*, Vol. 250, No. 4979 (Oct. 19, 1990), 353.

9   Ibid.

10 See also Philip H. Abelson, "Mineral Dusts and Radon in Uranium Mines," *Science*, Vol. 254, No. 5033 (Nov. 8, 1991), 777.

11 The number of annual lung cancer deaths has been variously estimated at between 130,000 and 140,000. The EPA's 1986 *Citizen's Guide* said that those attributable to radon exposure fell between 5,000 and 20,000. The 1992 edition said the range was between 7,000 and 30,000. U.S. Environmental Protection Agency, *A Citizen's Guide to Radon: What It Is and What to Do About It* (Washington, DC: U.S. Environmental Protection Agency, Aug. 6, 1986), 1; U.S. Environmental Protection Agency, *A Citizen's Guide to Radon: The Guide to Protecting Yourself and Your Family from Radon*, 2d ed. (Washington, DC: Government Printing Office, May 1992), 2.

12 Committee on the Biological Effects of Ionizing Radiations, National Research Council, *Health Effects of Exposure to Low Levels of Ionizing Radiation, BEIR V* (Washington, DC: National Academy Press, 1990).

CHAPTER 5

Federal Regulators and Congress

"RICH GUIMOND WAS, IS, AND I THINK ALWAYS WILL BE the driving force behind this country's radon policy," said a federal official in 1992. The official, who requested anonymity, was talking about the Environmental Protection Agency's (EPA) first radon division director, who has since risen to become the agency's assistant administrator of Superfund activities. The description of Guimond's powers may be exaggerated. But the former radon director himself acknowledges his pivotal influence on radon policy.[1]

Guimond began working on radon issues in the EPA during the 1970s, when few in the nation had heard of the gas. After the Watras incident in late 1984 and the subsequent discovery of elevated radon levels in other northeastern homes, the matter changed. Radon became a front-page issue.

At that point, Guimond recognized the opportunity for a far greater role both for the agency and himself. He met with Lee Thomas, the agency's administrator, and urged an aggressive EPA radon policy. He recalled that Thomas was reluctant at first because of the huge expense the government might face. "I knew the program would not be welcomed by OMB," Guimond said, referring to the White House Office of Management and Budget, which was perpetually concerned about keeping costs down. But he was able to convince Thomas to embark on a 5-year plan. "The fundamental part of the strategy would be that the money would not come from the federal government, but would mainly be a homeowner responsibility," Guimond said.

Soon after, Guimond was appointed director of the newly established radon division in the EPA's Office of Radiation Programs. One of his earliest efforts was to secure congressional backing for the EPA's plans. The goal was accomplished in 1986 with enactment of the first federal legislation on naturally occurring residential radon. The law designated the EPA as head of radon policy development and coordination.[2] Guimond does not hesitate to cite the importance of his role in the outcome. "We provided the strategies to the staffs on Capitol Hill, and we were pleased with the result," he says.

If Guimond has been at the core of the politics of radon, there are others in the government who have also enthusiastically joined in. This chapter recounts the roles of federal regulators, particularly environmental officials, and of members of Congress. Their participation in the development of radon policy is gleaned largely from hearings on the subject by congressional committees between 1985 and 1992. The hearings became a window through which development of an increasingly aggressive policy could be observed.

Federal Agencies

The EPA and the Department of Energy (DOE) have investigated radon since the 1970s. At the beginning of the 1980s, neither body was seen as more dominant than the other on radon-related issues. But the 1986 law codified the change in balance, and the EPA became chiefly responsible for developing radon programs. Since then, the agency's programs have become the measure against which groups and individuals have judged the appropriateness of the nation's policies.

Arguments about radon often turn on whether the EPA's policies have been too demanding or too permissive. But the agency's congressionally designated role has provided it with a platform that others do not enjoy. The EPA is the only body, public or private, whose spokespersons have been called to testify at all nine congressional hearings on the subject between 1985 and 1992. The hearings demonstrate the evolution in the agency's attitude, including increased certitude about the correctness of its policies.[3]

Congressional Testimony and the EPA

At a congressional committee hearing in October 1985, the first of nine held on radon during the next 7 years, the EPA largely reviewed its historical involvement with radon issues and its plans for the future. Sheldon Meyers, the agency's acting director of radiation programs, noted that during the previous 15 years the EPA had "developed a range of expertise and interests" as it assisted states with specific indoor radon problems.

Initially, Meyers testified, the EPA had focused on elevated radon exposure in structures built over uranium mill tailings, phosphate ores, and other wastes. Now it was assessing natural sources more carefully. The agency had established a Radon Action Program the previous month, in September 1985. Meyers explained that the nascent program would include epidemiological assessments, home surveys, and measurement and mitigation techniques.[4]

The published proceedings of the 1985 hearing included written answers by the EPA to questions submitted by Representative James H. Scheuer, who chaired the session. One pertained to the suggested action level of 4 picocuries of radon per liter of air. Since this number has often figured as a point of departure for criticism of the agency's radon policies, its genesis and rationale are of particular interest.

Scheuer's question was: "What is the basis for EPA's decision to use [4 picocuries of radon per liter of air] as the level at which mitigation is recommended?" The agency responded that the action level applied only to "technologically enhanced sources of radon," such as uranium mill tailings and other radium-contaminated sites. The EPA had "not made or proposed a decision" about 4 picocuries as an official action level for naturally occurring indoor radon. But the agency was clearly concerned. "Available evidence from studies of humans exposed to radon seems to indicate that risks are present for any exposure to radon and that the risk increases with exposure level and duration of exposure."[5]

These words doubtless inspired interest by environmental groups and others who called for an aggressive policy. But EPA also acknowledged that, as a matter of practice, low-level exposure was inevitable.

> In providing guidance to individuals on reducing radon exposures in their homes, the Agency is acutely aware of the need to provide reliable information so that homeowners will not spend money on unproven or unsuccessful techniques. Existing evidence supports the Agency's position that, in general, [4 pCi/L] is generally an achievable goal, whereas reductions below this level are often not practical and sometimes not even possible. Therefore, [4 pCi/L] is currently considered a practical lower bound for the general case of radon mitigation.[6]

Thus, the EPA in effect named the 4 picocurie figure the de facto action level because it seemed practical.

At a Senate hearing in April 1987, A. James Barnes, the agency's deputy administrator, proclaimed that since 1985, "we have adopted a national radon strategy and we have taken rather vigorous action."[7] He named four "major elements" in the EPA's radon program: problem assessment, mitigation and prevention activities, development of state and local government capabilities to work with homeowners, and public information. He outlined the agency's accomplishments in each of these areas, from instituting surveys in various states through dissemination of the *Citizen's Guide to Radon*, which was published in 1986. But questions by the senators suggested that several were still troubled by the EPA's 4 picocurie action figure.

Senator George Mitchell repeated the assumption that this level allows a cancer risk of 3 deaths out of 100 exposed. "Are you confident that the action level of 4 picocuries per liter is adequate to protect the public health?" Barnes responded that the technology to bring readings below that level was unavailable. But in the future "if we are able to do better than that, then we should do it."[8]

Barnes then made a startling observation about the ostensible danger of outdoor radon levels. "The risk that you and I run from radon in the outdoor air of about 1 in 1,000 is also higher than we would have under many of the regulatory programs that we have in place."[9] (Actually, the EPA's 1986 *Citizen's Guide to Radon* says the estimated number of lung cancer deaths from average outdoor radon levels is between 1 and 3 per 1,000.[10])

Barnes's statement raised several questions that were not pursued at the hearing. If the agency's conjecture were correct, a quarter million people in the current U.S. population would die of radon-induced lung cancer even if they lived a lifetime in homes with radon concentrations no greater than outdoors. He was saying as well that outdoor air that is otherwise uncontaminated was more risky than exposure to many other hazardous materials that were being regulated.

On the other hand, Barnes's supposition could be incorrect. Perhaps outdoor radon does not create the risks that he and the EPA assumed it does (which are based on linear extrapolations from the known effects of higher concentrations). If this is the case, then extrapolations about low indoor concentrations may also be faulty, and EPA's entire indoor radon policy would be questionable.

Another loose end that went unpursued occurred when Senator John H. Chafee asked Barnes what other nations are doing about radon. Barnes answered that Sweden was ahead of other countries on investigation and policy development and that "the Swedes have adopted a fairly long-term approach to dealing with this issue." Their goal was to reduce indoor levels by half in the next 40 or 50 years, Barnes said. Then he corrected himself: "I said 40 or 50 years? It is 100 years." Senator Chafee responded with mocked puzzlement: "That is what you call a 'gradual approach.' Let's not look to the Swedes for a solution. Let's look someplace else."[11] Evidently there was little interest in why Sweden, which Barnes acknowledged was ahead on the issue, had taken a less vigorous position. Chapter 9 returns to this question.

When Barnes appeared at a House committee hearing at the end of 1987, his prepared remarks paralleled those presented to the Senate committee. He emphasized again how much the EPA had accomplished during the past 2 years, mentioning the four-point program and noting that investigations were still under way. As in the Senate hearing, after Barnes completed his formal presentation he was asked about the adequacy of the EPA's action level. The issue dominated the question-and-answer portion of his testimony. Henry Waxman, chairman of the hearing, worried that the 4-picocurie figure was too high. "As I understand it, EPA has said that a 4 picocuries standard is about equivalent to 200 chest x-rays a year."[12]

Barnes's response showed his discomfort. As at the Senate hearing, he claimed that outdoor radon created a risk of 1 in 1,000. In effect, he was saying that little can be done about some risks and that indoor levels below 4 picocuries were among them. This led to a strained exchange between Waxman and Barnes. The following excerpts reveal the frustration that each felt about the other's position.

> *Waxman:* Why wouldn't you want to set regulations that insure the public's health? Wouldn't this be the way to give public assurances based on all the information you have, at least as of this point, that a certain level indoors is going to be adequate, and then we should try to achieve that level?
>
> *Barnes:* OK. Well, to begin with, Mr. Chairman, I have some problem with the statements that EPA has suggested this is a, quote, "safe level." If you look at the citizen's——
>
> *Waxman:* It's clear you haven't made it.
>
> *Barnes:* I mean we put out a citizen's booklet in which we showed what the risks are that flow from exposure at outdoor—average outdoor levels…. But I think we cause ourselves some other problems if we go out and tell the American people in strong terms, "The level of radon you're exposed to in the outdoor air poses an unacceptably high risk to you." At this point, there is nothing that we reasonably can tell them about what they can do to reduce it.
>
> *Waxman:* Well, no one's talking about the outdoor air. We're talking about the indoor air.
>
> *Barnes:* But, I mean, it's hard for me, Mr. Chairman, if you asked me what the Agency is likely to do if you give it the traditional charge to set…a level that's safe, within a margin of safety. I'd have to tell you that radon is a—it's a known carcinogen. In EPA's normal way of operating, we would set that goal at zero. That is a level that's below what's present in the outdoor air. It's not a level that's achievable….
>
> *Waxman:* We've never set a standard at zero. EPA has never set

a standard at zero.

> *Barnes:* Oh, but as I understand——
>
> *Waxman:* Excuse me. The point I'm trying to make—and I want you to listen because I want to make a point here, and then I want you to respond to it after you've listened to it—the point is this: You set a standard, you set a level that has been misread, as you acknowledge, to be the standard that would protect the public health. You acknowledge that the standard you set is not intended to be a health-based standard, that the public shouldn't think that, even if they achieve that level in their homes, they have achieved protection from the adverse consequences to their health. Now why wouldn't we have a standard set, based on whatever scientific information you have, so that people know what standard would protect their health in their homes, and then try to achieve that standard. If it can't be achieved immediately, at least we'd keep working on it....
>
> *Barnes:* As you have indicated, there is a residual risk when you come down below [4 picocuries]. At this point in time, the judgment of the professional community is, that's kind of the limit of where you're going to get with the technology that's there now....
>
> *Waxman:* But I think we need to set a goal, because I think the American people need to know that we are looking to achieve the protection of their health, at least in their own homes, from what is a serious carcinogen to which they are exposed.[13]

Six months later, in May 1988, Barnes appeared before a Senate committee that was inquiring into the manner that federal agencies were dealing with radon contamination. Once more he outlined the EPA's four-point program and claimed satisfaction with the agency's accomplishments. As before, questions about the action level were prominent. Senator Frank Lautenberg asked if the administration supported adoption of a federal radon standard that would require action and if the lack of a standard impeded efforts to address the radon issue. Barnes answered no to both questions, holding that the existing policy—an action level advisory—was quite appropriate. Invoking a formal standard that might be made mandatory was unnecessary, he said.[14]

As to the 4-picocurie figure, Barnes declared:

> Now, certainly, there is some debate. There are some people who say you ought to be concerned about lower levels, and there are a few people who would say maybe you could go to higher levels. But I would say the vast bulk of people would agree that at the level of 4 picocuries per liter of air, you should be concerned about it and, we believe, the current technology in most cases allows you to do something to reduce the exposure of individuals below that level.[15]

The momentum toward a more aggressive approach on the part of Congress resulted in the Indoor Radon Abatement Act in October

1988. The act began with the proclaimed long-term national goal to reduce all indoor radon concentrations to outdoor levels, with the effort to be overseen by the EPA. Around the time the law was enacted, the EPA and the Public Health Service had announced new evidence for the claim that 10 percent of all homes in the United States had radon levels above 4 picocuries. They now advised that every residence below the third floor be tested for radon.[16] Few home dwellers heeded the advice, and at the end of 1989, the EPA and the Ad Council began a campaign to induce Americans to test and mitigate.

A More Aggressive EPA

By the time hearings were held in the House and Senate in May 1990, national policy had become much more aggressive. The EPA was represented at the 1990 House hearing by F. Henry Habicht II, deputy administrator of the agency. His remarks reflected the striking turn in EPA's approach, not only in policy but tone. His praise for his agency's accomplishments was fulsome. "We are proud of the radon action program" that the agency began in 1985. "We are proud of the *Citizen's Guide to Radon*" that the agency published in 1986. As for the agency's measurement and mitigation program, "we are proud of that." The agency's radon people, in his judgment, were "doing an outstanding job."[17]

At a Senate hearing the following week, Richard Guimond spoke with similar certitude. He pronounced his agreement with the scientific assessments just made by Vernon Houk of the Centers for Disease Control. Houk had testified that "the evidence on radon is the strongest of any environmental contaminant," that "there is no room for debate on this issue," and that "anybody who tells you differently is ill-informed, deceitful, or both."[18]

Unlike at earlier hearings, not a trace of circumspection or frustration could be found in the words of EPA spokespersons. Was the new demeanor born of justified confidence, or was it masking insecurity about the wisdom of a policy they had embraced headlong? Whatever the reason, the questions by representatives and senators carried little of the bite that was evident at earlier hearings. None of the legislators questioned the appropriateness of the agency's action level.

Highlights from Habicht's 1990 testimony reflect both the heightened sense of importance of radon in national environmental policy and certitude by the EPA that its approach was correct. At the beginning of his written remarks, he said the agency considered "radon as the top environmental health problem facing the Agency and the nation." In his review of the agency's activities, he called good public information the "linchpin" of the radon program.[19] To that end, the agency was looking forward to the effects of the recently begun Ad Council campaign. Until now, Habicht noted, less than 3 percent of all homes had been tested despite the EPA's warnings.[20]

Representative Claudine Schneider wondered whether "in putting together your ad campaign [you] worked with any people who are familiar with what pushes people's buttons—you know, someone who is an expert in neurolinguistic programming or neurolinguistic conditioning, that sort of thing."[21]

Habicht responded: "Well, that is a very perceptive line of questions." He addressed her suggestion by describing an Ad Council effort along those lines.

> In developing this latest national campaign—which is both print and television and radio—about radon, it is much more hard-hitting than anything we had ever done before. As you recall—I know you are familiar with this—the television ad actually had a scene of a family that would flash x-ray-type pictures of the family to indicate that the radon exposure at certain levels is the equivalent of many hundreds of chest x-rays, as a way of grabbing people.[22]

Habicht said that some television stations refused to run the ad "because they thought that it was a little bit too hard-hitting." He guessed that "we don't have the message down perfectly" yet. But the EPA's basic thrust, as he affirmed earlier, was "to really grab people by the lapels on this issue."[23]

In testimony at a 1992 House hearing, Michael H. Shapiro, the agency's deputy assistant administrator for air and radiation, said the EPA was continuing to collect data to reduce uncertainty in radon risk estimates. While reiterating that indoor radon remains "one of the most important environmental health problems facing our country today," he proclaimed the EPA program "successful." Consistent with the agency's previous pronouncements, he praised the Ad Council campaign's "extensive contribution" to the effort.[24]

Shapiro further indicated that about six million homes have average annual radon levels that exceed 4 picocuries per liter. Unnoted was the fact that this was two million fewer such homes than the EPA had been claiming during the previous 5 years. Yet, strangely, he also said the number of radon-caused lung cancer deaths was now between 7,000 and 30,000, which, again unnoted, was *higher* than the agency's previously estimated range of 5,000 to 20,000. No one asked about the apparent contradiction.[25]

The Department of Energy

While the EPA was determined to shake the public from its lethargy, others were not as convinced that the public needed as much shaking. In 1985 John P. Millhone of the Department of Energy testified that since 1977 the department had been investigating radon measurement and control methods as well as its health effects. The department had determined that an action level of 8 picocuries per liter of air was appropriate (twice the level advocated by the EPA) and that perhaps one million homes might need remediation. This was several million fewer than would presumably require remediation at the EPA's 4-picocurie action level. The emphasis of his testimony was on the need for more knowledge rather than an aggressive policy.

> While much progress has been made in improving our understanding of radon and its effect on health, additional work needs to be done in several areas. These concern the development of additional data on radon concentrations and anomalies, the assessment of the health risks associated with these findings, and the effects of energy conservation activities and various mitigation strategies on indoor radon levels.[26]

In written responses to questions raised at the hearing, the DOE and EPA made clear that they were working cooperatively on radon. Thus in 1985, the DOE indicated that the two agencies "regularly discuss planned and current radon research" and "jointly plan and conduct radon research."[27]

Similarly, the EPA indicated that it respected the DOE's "long-standing radon program" and that it coordinated its research with DOE by way of interagency meetings. Moreover, the two agencies regularly communicated "informally through the close lines of communication that have developed among staff." Their frequent interagency contacts helped "ensure that the Federal projects are complementary and not duplicative."[28]

In 1985, therefore, understandings about radon policies by the two agencies seemed in harmony. Indeed, EPA was respectful of DOE's long-standing work and knowledge in the radon area. The mission of the two agencies differed, however. The DOE's mission was to engage in basic research, while EPA's was to focus on applied work and operational programs.[29]

After 1985, although EPA spokespersons testified at every congressional hearing, DOE representatives were rarely invited. Congress was giving EPA increased responsibility for radon policies, and information developed by DOE seemed of less interest on Capitol Hill. As the decade unfolded and EPA's views were expressed with greater certitude, DOE's approach remained circumspect. In 1988, Susan Rose, the department's radon program manager, implicitly criticized EPA's ambitious approach. She suggested that information on the risk of radon was "uncertain enough" not to warrant a policy of widespread testing.[30]

In 1989, a DOE report again underscored the uncertainty of the matter. "Significant scientific questions yet remain about extrapolating the lung cancer risk from radon exposure in uranium miners, the source for radon risk estimates, to that of the general public."[31]

A 1992 DOE memorandum criticized EPA for pressing Congress to pass more radon legislation. It called a bill to reauthorize the 1988 Indoor Radon Abatement Act "a 'budget buster' of private sector costs." The 1988 act, according to the memorandum, "was a costly program-building bill that did not provide *any* (original emphasis) health benefits. It did provide the EPA program and the radon industry growth opportunities."[32]

The infelicitous appraisal indicated how far cooperation had given way to disagreement between the DOE and the EPA. Perhaps, as was suggested at the 1992 House committee hearing, the dispute was largely based on jealousy and competition for power. At the same time, the hearing demonstrated the consequences of failure to invite DOE representatives to testify.

During the hearing, chairman Al Swift asked Michael Shapiro, the EPA witness, what the DOE thought about the radon bill under consideration. After a long pause, Shapiro said he could not speak for the DOE. An aide whispered into chairman Swift's ear for several seconds as the hearing room remained silent. The chairman then noted with a smile that apparently there were "turf" differences between the EPA and the DOE, and turned to other matters.[33] The department's views went unheard.

The DOE later was asked to submit a written statement for incorporation into the record of the hearing. But the opportunity for dialogue, for members of Congress and the press to hear the department's rationale in person, was again lost. As at other congressional hearings, failure to invite DOE representatives meant failure to hear another informed range of opinions on the nation's radon policy.

Department of Housing and Urban Development

The only other federal agency to provide substantial testimony was the Department of Housing and Urban Development (HUD). It was summoned to respond to a report issued by the General Accounting Office (GAO) about the government's response to dealing with radon in federally financed housing.[34] The report was requested by Senator Frank Lautenberg, who expressed concern to the GAO that federal agencies might not be acting adequately to reduce radon "contamination" in homes. The report cited four agencies, all of which had shown "a limited radon response." But the GAO minimized the significance of the role of three agencies—the Farmers Home Administration, the Veterans Administration, and the National Park Service. The fourth, HUD, was the only one, along with EPA, that had a "statutory mandate to address indoor radon hazards." The report criticized HUD's actions as "limited" and "piecemeal."[35]

HUD spokespersons were criticized more directly than officials from any other agency. Their congressional interlocutors were not pleased with the GAO finding. In May 1988, Jack R. Stovkis testified on behalf of HUD that "the Department disagrees with the overall GAO finding" and that there were "serious errors of omission and fact throughout the document." He argued that the GAO failed to take into account actions by HUD, including its coordination with other federal agencies on radon research. He disputed the GAO's contention that the housing department had no official radon policy, insisting that its "policy is to provide assistance on radon matters at the request of State or local government officials."[36]

Senator Lautenberg became impatient with Stovkis, as excerpts from his interrogation reveal.

Lautenberg: Mr. Stovkis, why should the Nation's housing agency knowingly sell homes that are defective homes with elevated radon risks?

Stovkis: Why should we knowingly? The only way we knowingly do that is with a full public disclosure of the health hazard.

Lautenberg: Then why shouldn't you get on with the task of identifying the risks in whatever way that you can? You have a responsibility to protect those lives. What I hear as we go through this testimony is that there are more excuses than reasons for not getting on with this, whether we are talking about authority or funding or relationships with EPA or what have you.

I am not satisfied with the answers that we are getting here today.[37]

Lautenberg warned that "lives are at stake" and admonished Stovkis to "get your agency on the ball." He then asked how Congress might help with further legislation. Stovkis was unenthusiastic about Lautenberg's offer. He worried about "laws that require Federal agencies to spend a lot of money and then create a lot of unintended problems. I think probably one of the best ways to deal with radon is by self testing." He acknowledged HUD's responsibility to help educate the public about radon, but demurred from endorsing a more aggressive role.[38]

In May 1990, a House committee conducted hearings on federal efforts to promote radon testing. John C. Weicher, HUD's Assistant Secretary for Policy, Planning and Development, was treated no more gently than Stovkis had been 2 years earlier. Weicher's statement reviewed recent legislation that mandated increased HUD involvement with radon policy. The 1986 radon law required HUD to work with EPA to assess and reduce indoor radon concentrations in new construction.[39]

Beyond that, two laws in 1988 added to HUD's responsibilities. The Indoor Radon Abatement Act required that EPA develop model construction standards for controlling radon levels in new buildings.[40] Although not designated by name in the act, HUD participated with the National Institute of Building Sciences in helping EPA's efforts. Another piece of legislation required HUD to develop policy to deal with radon contamination in public housing.[41]

Weicher's statement seemed to give little comfort to his congressional interlocutors. Representative James Scheuer asked why HUD did not routinely seek radon tests on all sales or resales of federally insured mortgages. Would not that be the best way to alert people to the radon problem, he asked. Weicher barely uttered, "That is a possibility," when Scheuer interrupted: "Wait a minute. Wait a minute. I'm not interested in hearing about possibilities." The congressman admonished Weicher, saying that "[y]ou require a hell of a lot of other things when a property is being sold…. Why isn't it on the top of your priorities list?"

Weicher suggested that reliable testing would take more than a week, particularly in homes that have been repossessed and vacant for substantial periods.[42] His response did not satisfy Scheuer, who continued to urge action along the lines advocated by the EPA.

As is evident from this overview, at the end of nearly 7 years of radon policy, EPA dominance in the matter remained solid. Initially, the most visible spokespersons on behalf of a more aggressive policy were not agency officials, but a few members of Congress. Occasionally pressing the agency to do even more, they made the EPA appear as a moderate force, fending off extremes. By the end of the decade, however, the agency had adopted outwardly aggressive policies.

Congress on the Offensive

The first declaration in 1986 by the EPA and Surgeon General that indoor radon was a widespread problem had come on the heels of several environmental setbacks for the administration. During Ronald

Reagan's first presidential term, environmental policies of his otherwise popular presidency were in constant trouble. Several EPA officials, including Anne Gorsuch, the agency's head, were forced to resign in 1983 in the face of charges that they were not enforcing environmental laws. Soon after, James Watt left his position as Secretary of Interior, following criticism that he placed business interests over the environment. The administration's attempt to weaken environmental laws during its first term defied public preferences to strengthen them.[43]

Radon seemed an ideal issue for a conservative administration in need of a better environmental image. The radioactive gas was business friendly insofar as it is a natural phenomenon and not usually related to industrial activities. Since testing and remediating would be at the homeowners' expense, the cost to government would be negligible. Yet by proclaiming radon a major health hazard to the general citizenry, the government was demonstrating concern about environmental health and safety.

"Some of the attention radon has been receiving," *New York Times* reporter Philip Shabecoff wrote in 1986, "may also be explained by the fact that it is an environmental problem that the Reagan administration can meet without having to spend a great deal of Federal funds or to impose new rules and costs on industry."[44] Another newspaper writer later privately described the administration's policy to me as "environmentalism on the cheap." Whether or not the administration's initial position was rooted in partisan politics, the question soon became moot. Congressional representatives from both parties began to take interest in the matter. Indeed, within a few years the most dire statements were coming from Democrats on Capitol Hill. In the early 1990s, media attention to radon had declined and the public was scarcely interested. But senators and representatives, mostly Democrats, were introducing more radon bills than ever.

Members of the House

In a 1991 hearing on indoor air pollution by the House Subcommittee of Health and the Environment, Democratic Congressman Edward Markey announced that he was reintroducing legislation "to tackle radon head on." He called the gas a "silent killer" that is responsible for "a public health disaster."[45] He repeated his warning in 1992, adding that "even as we speak, *10 million school children* in this country are receiving the cancer threat of smoking *half a pack of cigarettes a day*" (original emphasis was in transcript distributed at the

1992 hearing).[46] His bill would require certification of radon testers, testing of every school in the nation, testing of all houses seeking federally subsidized mortgages, and creation of a President's Commission on Radon Awareness.

Rivaling Markey's rhetoric at the 1991 hearing, Congressman James H. Scheuer declared that radon is among "the most important environmental and public health issues of our time." He reminded the committee that he had cosponsored radon bills in previous sessions of Congress and now was cosponsoring an indoor air quality bill with a fellow Democrat, Joseph P. Kennedy.[47] These congressmen were joined in their hyperbolic statements by Henry A. Waxman, the subcommittee chairman who in previous years presided over other hearings on the radon issue.

In 1991, as before, Waxman spoke for mandatory testing of homes. He characterized radon as "a frightening problem" and expressed dissatisfaction when F. Henry Habicht II, the EPA's deputy administrator, said, "We don't think society or we are ready for a mandatory program."[48]

As described above, Waxman had shown similar impatience with the EPA in 1987. At that time, he questioned whether the 4-picocurie action level was too high and urged a lower "safe" level even if it seemed unattainable. "If it can't be achieved immediately, at least we'd keep working at it."[49] This attitude formed the provision of the 1988 Indoor Radon Abatement Act that calls for reducing indoor concentrations to ambient outdoor levels.

Waxman and his subcommittee staff base their assessments on the supposition that no amount of radiation can be considered safe. Thus, natural outdoor levels must also be considered hazardous. This explains why Gregory S. Wetstone, counsel to Waxman's subcommittee, could imply that the 1988 act did not go far enough.

> The Act doesn't even say reduce radon to a safe level, but only to the ambient level. To make the objective a reduction to the ambient level is a different thing from reducing it to a safe level, like zero or close to that.[50]

Once an enclosure is created on the earth's surface, radon is likely to seep in and not dissipate as it would outdoors. Wetstone does not believe that reducing indoor concentrations to outdoor levels would be difficult to achieve, however. Despite misgivings by scientists who think the task formidable, he says "I don't think it's unrealistic. If you have a ventilation system you should be able to get close to that."

Wetstone, who worked on the development of the 1988 act, said that basically, the ambient-level provision was a goal. The intent was to get away from the 4-picocurie level which EPA then persisted with anyway. 4 picocuries is misleading. People now think anything below 4 is safe. If you buy a home and it is 3.5 picocuries, people think that is within EPA's safe limits.... EPA seems to imply that you don't have a concern if your level is below 4.[51]

What should EPA's policy be? Wetstone says it should be to "get the levels as low as possible. People should be told they can get the levels significantly below 4 and that this is worth doing." The committee's chairman, Congressman Waxman, "feels the same way," says Wetstone. Indeed, the congressman made this sentiment clear during comments at his subcommittee's hearings.

Not all congressional radon policy enthusiasts are Democrats. In 1990, Claudine Schneider, among other House Republicans, said she considered indoor radon a "serious health threat."[52] But the few congressional representatives who expressed skepticism about the severity of the radon problem were Republican conservatives. Their questions largely focused on the mechanics, rather than validity of policy, and on the more hyperbolic phrases by advocates of aggressive action. Thomas J. Bliley, Jr., of Virginia, for example, wondered whether EPA, DOE, and the Occupational Safety and Health Administration were adequately coordinating their radon activities.[53] And he asked for documentation of a statement by Marvin Goldstein of the American Association of Radon Scientists and Technologists that "radon is the most important carcinogen to be dealt with in this century."[54] (Goldstein cited as his source remarks by Richard Guimond at a radon conference in 1990.[55])

Nor were skeptics above sarcasm. William E. Dannemeyer of California, an extreme conservative, ridiculed the issue during a hearing on legislation to assist states in developing radon policies.

I'm not sure I understand how we can, by law, stop the Earth from emitting a substance that some say is hazardous to health. It just boggles my imagination. But as I say, I've never ceased to be amazed by the ingenuity of my colleagues on the left, running the House of Representatives.[56]

The issue of course had nothing to do with stopping the earth from emitting radon, but rather how to respond to a possibly hazardous situation. But such rhetoric from the right was rare. For the most part the belief that radon posed a major health hazard to the public went unchallenged in the House.[57]

By the early 1990s, radon legislation was unmistakably being driven by Democrats, usually liberals. This was true in both houses of

Congress, as Senate leapers proved no less aggressive than those in the House.

Members of the Senate

In May 1991, the Senate Subcommittee on Superfund, Ocean, and Water Protection held a hearing on four radon bills. Collectively, the bills would require the EPA to measure radon in schools throughout the country and help provide funds for remediation, to establish a procedure by which all home buyers receive information about radon, to impose mandatory testing prior to sale of homes involving federal financing, and to require certification of testers.[58]

As in the House, the thrust for radon legislation largely came from the Democrats. Two of the three senators who introduced or cosponsored the bills were Democrats, Frank Lautenberg and George Mitchell, and the third a Republican, John Chafee. Chafee made clear his difference with the Bush administration. The administration not only opposed any mandatory testing but also a requirement that home buyers be informed about radon. "I must confess to being somewhat disappointed" by the administration's lack of support, Chafee said.[59] The Democratic-environmentalist surge toward mandated action against indoor radon was increasingly challenging the Reagan and Bush approaches to urge people to act on their own.

Frank Lautenberg

No legislator has been more dedicated to the development of an aggressive radon policy than Senator Frank Lautenberg of New Jersey. He has introduced or cosponsored most Senate bills on the subject, including the two that became law in the 1980s: the 1986 directive that the EPA develop more information about mitigation methods and the location and severity of radon concentrations throughout the country; and the 1988 Indoor Radon Abatement Act that expanded the government's involvement with surveys, testing, and research on radon. He has also championed bills to fund states to develop radon policies, train health officials in dealing with radon, encourage testing of houses and public buildings, and allow for tax deductions of testing and mitigation as medical expenses.

As chairman of the Subcommittee on Superfund, Ocean, and Water Protection (and previously the Subcommittee on Superfund and Environmental Oversight), he has presided at hearings on the subject most years since radon became a national issue. In 1987, while calling radon "a major public threat," he decried the administration's "lack of commitment" to a more vigorous policy.[60]

In 1988, at a hearing on how federal agencies were dealing with radon "contamination," he expressed concern about "what Federal agencies are doing or, more accurately, not doing about radon despite the fact that unsafe radon levels threaten over 10 million homes in America; despite the fact that radon causes as many as 20,000 lung cancer deaths a year, second only as a threat to cigarette smoking."[61] "Frustrating as the devil for me" is how Lautenberg described the Department of Housing and Urban Development's failure to address radon and promote "safe housing."[62]

In 1990, his subcommittee heard testimony on a bill to require radon testing in schools. Lautenberg cited an EPA survey that found that more than half the schools tested had at least one classroom with radon levels above 4 picocuries. Said Lautenberg, "Our Nation's children are needlessly exposed to dangerous radon levels," and further inaction is "inexcusable."[63]

In 1991, Lautenberg introduced the reauthorization bill for the 1988 Indoor Radon Abatement Act. The bill, which is discussed below, expanded the EPA's scope of activities, including the designation of "priority radon areas," further promotion of public awareness "about radon health risks," and action to reduce radon levels. The national goal of reducing indoor concentrations to outdoor ambient levels was left undisturbed.[64]

At the subcommittee hearing on the bill and other proposed radon legislation, Lautenberg noted the minimal public response thus far to the radon issue—that only 5 percent of homes had been tested. "We've got to expand the efforts" to encourage people to test and remediate, he said. He reiterated his frustration with HUD, saying that other agencies were more responsive, and the department had yet to "begin the process of stopping this killer."[65]

Lautenberg is a silver-haired former businessman. He is comfortably voluble and determined to deal with radon and other environmental issues. After election to the Senate in 1982, he successfully sought membership on the Environment and Public Works Committee where he could best influence national environmental policy.

The senator has led efforts to reduce air and water pollution, and he has shown special sensitivity on the subject of cancer. He has introduced legislation not only on radon, but on other carcinogens such as asbestos and tobacco smoke. He was the author, notably, of legislation prohibiting smoking on airplanes during flights within the United States.

Lautenberg's lifetime experiences have deeply affected his legislative focus. He freely notes that his father died of cancer and that he wanted to "do what I can to eliminate radon as a cause of cancer."[66] An aide recalls that Lautenberg's friend also died of cancer, allegedly caused by exposure to asbestos. Along with his father's death, said the aide, this has strongly influenced his personal and legislative agendas.[67]

Robert Ferguson, executive vice president of the New Jersey Association of Realtors, has had many dealings with Lautenberg since his election to the Senate. He believes the matter of controlling cancer is an obsession with the senator.[68]

Lautenberg's biography, provided by his office, reveals his unusual concern with the subject. The six-page document notes that he is "determined to help prevent the scourge of cancer that had taken his father's life prematurely." The final sentence of the biography refers to his activity on behalf of Soviets who wish to emigrate to the United States and "who wish to leave to seek treatment for cancer in the West."

Ironically, Lautenberg is a resident of Montclair, the New Jersey town that underwent the radon policy contretemps discussed in Chapter 6. The Montclair problem first became known to the public late in 1983, during Lautenberg's first year in office. Not until 1985, after the Watras incident occurred in Pennsylvania, did radon become an issue of national concern. But after his town's experience, radon was a familiar word to the senator. Thus, the ingredients for his special interest in radon were in place by the time the gas claimed national attention.

In a highly publicized effort to gain information on the subject, Lautenberg led a delegation of Americans to Sweden in January 1986. He has referred to that trip repeatedly during congressional hearings as important to the development of his understanding of proper policy. Yet paradoxically, he has revealed inconsistencies, if not misunderstandings, about some of his experiences there. Swedish policy and the views of Swedish experts about American policy are reviewed in Chapter 9. But a sample of Lautenberg's references to his Swedish experience makes the point.

At the 1987 hearing Lautenberg noted that he was accompanied on the trip by representatives of the building industry and Richard Guimond, director of radon programs for the EPA. He spoke admiringly about the Swedish program where "the threat was recognized at a much earlier stage" than in the United States. The Swedish national government, he said, "insisted that municipalities pick up the cost for testing because they didn't want that argument out there going on. What they wanted to do was find out whether people were exposed and get done with it."[69] Lautenberg's observation suggests that Swedish policy has been more directive and has gained more public compliance than in the United States. As the chapter on Swedish and Finnish policy demonstrates, that has not been the case.

At a hearing the following year, Lautenberg offered a similar theme.

> I visited Sweden and went there specifically to see how they deal with radon. They treat it as a matter of course, but they treat it in such a way that the person who is going to inhabit a home is alerted to it, that the home is required to have certain conditions met.[70]

His implication is that homes in Sweden are required by law to have radon tests and, if concentrations are at certain levels, undergo remediation. This is not the case.

Again, in 1991, Lautenberg referred to his experience in Sweden where

> the whole community is conscious of radon and there are very specific rules that you have to mitigate against it. If I remember correctly, you can't dispose of a piece of property without having said this is a radon-free piece of property.... Everybody does it, it becomes a relatively small problem.[71]

His view of the success of Sweden's radon program and public compliance is matched neither by the reality of the Swedish experience nor by the views of the country's radon experts.

Lautenberg has been as prone as any to use colorful words, like "killer," to describe radon. But inflated rhetoric and suggestions that Sweden's radon policies are more aggressive than they are do little service to developing a rational science policy. In fact, radon policies in Sweden and Finland have been more temperate than those of the United States and, given the present level of knowledge, have been quite rational.

The Indoor Radon Abatement Reauthorization Act

Lautenberg's leadership on the radon question was never more evident than in March 1992, when the Senate voted to reauthorize an amended Indoor Radon Abatement Act. Prior to the vote, he initiated debate on the Senate floor by proclaiming that the bill would strengthen the testing, mitigation, and education programs in the 1988 law. "Radon," he said, "is a known killer. It attacks us in our homes, our schools, and our work places.... The evidence is overwhelming."[72]

The bill provided for another $31 million in federal expenditures. But it included a new provision that would vastly increase the number of homeowners who would have to pay out of their own pockets for mitigation. The bill called for the EPA administrator to designate "radon priority areas." These areas are described in the bill as areas where "the average indoor radon level...is likely to exceed the national average indoor radon level by more than a de minimis amount."[73]

The average indoor concentration is about 1.3 picocuries per liter, well below the existing action level of 4 picocuries. By the EPA's account, the new provision would mean "about one-third of the country would be considered high risk."[74] An interpretation by a DOE analyst said the bill would cost homeowners more than $50 billion and make "two-thirds of the *entire* (original emphasis) U.S. victim to this Act."[75] None of these questions was discussed on the Senate floor.

A few Republicans voiced skepticism about the bill, and the Bush administration said it was costly and unnecessary. But senators from both parties spoke in support. Several, including Republican David Durenberger and Democrat Harris Wofford, commended Lautenberg for his leadership on the issue. Another Democrat, Quentin Burdick, praised him "for crafting this vital legislation."[76] The vote was 82 in favor, 6 opposed, and 12 not voting.

A House subcommittee held a hearing on a companion bill in June. In September, the House passed an altered form of the bill by voice vote. Less expansive than the Senate bill, it would require the EPA to establish mandatory standards for radon technicians and testing equipment. It also would have the EPA devise a strategy to identify areas and buildings "with exceptionally high levels of radon." And it would establish a presidential commission to increase "public awareness of the risks of exposure to radon."[77]

Only one congressman expressed hesitation about the bill during the floor debate. Don Ritter, a Republican from Pennsylvania, alluded to uncertainty about risk from exposure to low levels of radon. Although he said he would support the bill, he wondered whether mitigation at low levels was "an unnecessary use of scarce resources."[78] No one else even expressed a doubt. More common were unexamined pronouncements like that of Democratic Congressman Al Swift. "Radon kills more people than do airplane crashes, drowning, and fires—added together."[79]

The 1992 session of Congress ended without reconciling the differences between the bills passed by the Senate and the House. But the momentum for action assured that the issue would not disappear. The efforts by Lautenberg, Waxman, Markey, and others show how effective singularly dedicated lawmakers can be.

To dissect Lautenberg's approach to radon is not to detract from his commitment to promoting public health, especially in the area of cancer. But he has probably been the most influential member of Congress on the issue. The basis of his positions, the accuracy of his assumptions, and the consequences of his leadership therefore deserve scrutiny. Along with a handful of other senators and representatives, he has driven federal radon legislation. Each new bill has given EPA more power and incentive to become increasingly aggressive in its radon policy.

Some scholars emphasize that on a particular issue, a federal agency may frame the agenda for Congress. "If the agency does not pose the right questions, those questions are not likely to emerge in the course of congressional consideration."[80] But however much Richard Guimond or other EPA officials may have influenced the radon agenda, spokespersons with alternative views were readily available to the lawmakers. If senators and representatives have heard them infrequently, that has been their choice and that of their staffs. In the end, only Congress is responsible for the laws it enacts.

Notes

1   Unless otherwise noted, comments attributed to Richard Guimond are from an interview, June 11, 1992.

2   Radon Gas and Indoor Air Quality Research Act of 1986, Title IV, Pub. L. No. 99-499, §403, Oct. 17, 1986.

3   The EPA was established in 1970 as an executive branch agency. As a regulatory agency, it deals with both public health and resource management, though its administrator reports to the President. For a critical history, see Marc K. Landy, Marc J. Roberts, and Stephen R. Thomas, *The Environmental Protection Agency: Asking the Wrong Questions* (New York: Oxford University Press, 1990).

4   House Subcomm. on Natural Resources, Agricultural Research and Environment of the Comm. on Science and Technology, *Hearing on Radon and Indoor Air Pollution*, Oct. 10, 1985 (Washington, DC: Government Printing Office, 1986), 10–85, 114.

5   Ibid., 274

6   Ibid., 274–75.

7   Senate Subcomms. on Environmental Protection, and Superfund and Environmental Oversight of the Comm. on Environment and Public Works, *Hearing on Radon Gas Issues*, Apr. 2, 1987 (Washington, DC: Government Printing Office, 1987), 13.

8   Ibid., 17.

9   Ibid.

10 U.S. Environmental Protection Agency, *A Citizen's Guide to Radon: What It Is and What to Do About It* (Washington, DC: U.S. Environmental Protection Agency, 1986), 10.

11 *Hearing on Radon Gas Issues*, Apr. 2, 1987, 20.

12 House Subcomm. on Health and the Environment of the Comm. on Energy and Commerce, *Hearing on Radon Exposure: Human Health Threat*, Nov. 5, 1987 (Washington, DC: Government Printing Office, 1988), 56.

13 Ibid., 56–58.

14 Senate Subcomm. on Superfund and Environmental Oversight of the Comm. on Environment and Public Works, *Hearing on Radon Contamination: How Federal Agencies Deal with It*, May 18, 1988 (Washington, DC: Government Printing Office, 1988), 15.

15 Ibid., 16.

16 Philip Shabecoff, "A Major Radon Peril Is Declared by U.S. in Call for Tests," *New York Times*, 13 Sept. 1988, A-1.

17 House Subcomm. on Natural Resources, Agriculture Research and Environment of the Comm. on Science, Space, and Technology, *Hearing on Federal Efforts to Promote Radon Testing*, May 16, 1990 (Washington, DC: Government Printing Office, 1990), 4–6.

18 Senate Subcomm. on Superfund, Ocean, and Water Protection of the Comm. on Environment and Public Works, *Hearing on Radon Testing for Safe Schools Act*, May 23, 1990 (Washington, DC: Government Printing Office, 1990), 10.

19 *Hearing on Federal Efforts to Promote Radon Testing*, May 16, 1990, 5–7.

20 Ibid., 40.

21 Ibid., 41.

22 Ibid.

23 Ibid., 5, 41.

24 House Subcomm. on Transportation and Hazardous Materials of the Comm. on Energy and Commerce, *Hearing on Radon Awareness and Disclosure*, June 3, 1992 (Washington, DC: Government Printing Office, 1992), 30, 34 (testimony by Michael H. Shapiro, EPA deputy assistant administrator for Air and Radiation).

25 Ibid.

26 *Hearing on Radon and Indoor Air Pollution*, Oct. 10, 1985, 134–35.

27 Ibid., 254.

28 Ibid., 269–70.

29 In an interview on June 9, 1992, Margo Oge, head of EPA's Office of Radiation Programs, described the difference somewhat less generously. "Our mission is concerned with health and environment, and the DOE's is evaluating uncertainties. We are trying to control the problem, and they are trying to study it."

30 "Radon Home-Test Warning Attacked," *Record* (Hackensack, NJ), 3 Oct. 1988, A-10.

31 U.S. Department of Energy, Office of Health and Environmental Research, and Commission of European Communities, Radiation Protection Programme, *International Workshop on Residential Radon Epidemiology* (Washington, DC: U.S. Dept. of Energy, July 1989), foreword.

32 U.S. Department of Energy, Draft Memorandum, "Comparison of Indoor Radon Abatement Act of 1988 vs. Indoor Radon Abatement Act of 1992," May 20, 1992. (Excerpts from the DOE memorandum are in Appendix D.)

33 *Hearing on Radon Awareness and Disclosure*, June 3, 1992, 37. Characterizations are based on my observations during the proceedings.

34 General Accounting Office, *Indoor Radon: Limited Federal Response to Reduce Contamination in Housing* (Washington, DC: Government Printing Office, Apr. 1988).

35 Ibid., 3–5.

36 *Hearing on Radon Contamination: How Federal Agencies Deal with It*, May 18, 1988, 11–12.

37 Ibid., 32.

38 Ibid., 33.

39 Radon Gas and Indoor Air Quality Research Act of 1986, Title IV, Pub. L. No. 99-499, §403, Oct. 17, 1986.

40 Indoor Radon Abatement, Amendment of the Toxic Substances Control Act, Title III, Pub. L. No. 100-551, §744, Oct. 28, 1988.

41 Stewart B. McKinney Homeless Assistance Amendments Act of 1988, Title X, Pub. L. No. 100–628, cited at *Hearing on Federal Efforts to Promote Radon Testing*, May 16, 1990, 29–31.

42 Ibid., 43–44.

43 Norman J. Vig and Michael E. Kraft, eds., *Environmental Policy in the 1980s: Reagan's New Agenda* (Washington, DC: CQ Press, 1984), ix–x.

44 Philip Shabecoff, "Issue of Radon: New Focus on Ecology," *New York Times*, 10 Sept. 1986, A-24.

45 House Subcomm. on Health and the Environment of the Comm. on Energy and Commerce, *Hearing on Indoor Air Pollution*, Apr. 10, 1991 (Washington, DC: Government Printing Office, 1991), 3–4.

46 *Hearing on Radon Awareness and Disclosure* , June 3, 1992, 24.

47 *Hearing on Indoor Air Pollution*, Apr. 10, 1991, 4.

48 Ibid., 53.

49 *Hearing on Radon Exposure: Human Health Threat*, Nov. 5, 1987, 56–57.

50 Interview, Jan. 8, 1992.

51 Ibid.

52 House Subcomm. on Natural Resources, Agriculture Research and Environment of the Comm. on Science, Space, and Technology, *Hearing on Federal Efforts to Promote Radon Testing*, May 16, 1990 (Washington, DC: Government Printing Office, 1990), 42.

53 *Hearing on Indoor Air Pollution*, Apr. 10. 1991, 56.

54 Ibid., 262.

55 Ibid.

56 *Hearing on Radon Exposure: Human Health Threat*, Nov. 5, 1987, 59.

57 In 1992, Representative Don Ritter, a Republican conservative with an engineering background, said he had concerns about the risk assessments behind EPA's "radon action program." He hoped to hear more from scientists "who have raised questions concerning the effects of radon at low exposures." *Hearing on the Radon Awareness and Disclosure Act*, June 3, 1992.

58 Senate Subcomm. on Superfund, Ocean, and Water Protection of the Comm. on Environment and Public Works, *Hearing on Pending Radon and Indoor Legislation*, May 8, 1991 (Washington, DC: Government Printing Office, 1991) (S. 575, S. 791, S. 792, and S. 779).

59 Ibid., 14.

60 Full citation previously given in note 7. *Hearing on Radon Gas Issues*, Apr. 2, 1987, 3.

61 Senate Subcomm. on Superfund and Environmental Oversight of the Comm. on Environment and Public Works, *Hearing on Radon Contamination: How Federal Agencies Deal with It*, May 18, 1988 (Washington, DC: Government Printing Office, 1988), 1.

62 Ibid., 32.

63 Full citation previously given in note 18. *Hearing on Radon Testing for Safe Schools Act*, May 23, 1990, 1.

64 S. 792, A Bill "To Reauthorize the Indoor Radon Abatement Act of 1988 and for Other Purposes," Apr. 9, 1991.
65 *Hearing on Pending Radon and Indoor Air Legislation*, May 8, 1991, 1–2.
66 Personal communication, Feb. 25, 1992.
67 Rick Erdheim, interview, Jan. 7, 1992.
68 Interview, June 12, 1991.
69 *Hearing on Radon Gas Issues*, Apr. 2, 1987, 23–24.
70 *Hearing on Radon Contamination: How Federal Agencies Deal with It*, May 18, 1988, 32–33.
71 *Hearing on Pending Radon and Indoor Air Legislation*, May 8, 1991, 16.
72 *Congressional Rec.* S2960 (March 10, 1992).
73 Ibid., S2995.
74 *Hearing on Pending Radon and Indoor Air Legislation*, May 8, 1991, 43.
75 U.S. Department of Energy, Memorandum, May 20, 1992. Location in Appendix was given in note 32.
76 *Congressional Rec.* S2962 (March 10, 1992).
77 *Congressional Rec.* H9698-99 (Sept. 29, 1992).
78 Ibid., H9702.
79 Ibid., H9701.
80 Landy, Roberts, and Thomas, 4.

CHAPTER 6

Radon in the States and the Case of New Jersey

ALTHOUGH PRIMARY RESPONSIBILITY FOR THE NATION'S radon policy lies with the Environmental Protection Agency (EPA), federal legislation emphasizes the need for action at the state level. More than half the 10-page 1988 Indoor Radon Abatement Act relates to activities to be taken by the states under the auspices of the EPA. The law requires the EPA administrator or his designee to "develop and implement activities designed to assist state radon programs." The activities include provision of public information materials, surveys of radon levels, measurement and mitigation methods, and training seminars for state and local officials and private firms dealing with radon.[1]

In addition to the required activities, the law provides for "discretionary assistance." Upon request from a state, the EPA administrator or his designee may help design and implement surveys to locate radon in the state, programs of public information and education, and programs to control radon in new and existing structures. For these and other efforts, the law authorized up to $10 million a year in grant assistance to the states.[2]

Despite the programmatic and financial incentives, and dire warnings by federal officials about the danger of indoor radon, most states have been minimally responsive. By the end of 1989, after the EPA/Advertising Council's aggressive and widespread campaign had begun, the EPA listed only 13 states as having undertaken statewide radon studies. The EPA's list was dated October 16, 1989. In March 1990, I wrote to the designated contact person for each state on the list and asked about the status of the radon program in the state. I received replies from officials in nine states: Florida, Idaho, Louisiana,

Montana, New Hampshire, New Jersey, North Carolina, Oregon, and Washington. Except for North Carolina, all had begun (and in some cases completed) their intra-state studies before the 1988 law was enacted.

Among the four states that did not respond, two were known to have begun studies before 1988—New York and Pennsylvania.[3] (The other two were Virginia and South Carolina.) Thus, while some states may have been enticed to do more, the legislation seemed to have little immediate effect. Officials expressed no sense of urgency in their replies to me. Joanne Mitten of the Idaho Department of Health and Welfare wrote in her letter of March 26, 1990, that a statewide radon survey was under way, but "at this time we do not have the resources to comprehensively evaluate specific geographical areas." Jay Mason of the Louisiana Nuclear Energy Division noted on March 22 that his state's radon program is "minimal." "We mainly provide information on the topic. We have no legislation governing radon testing or mitigation." Adrian C. Howe, Chief of the Montana Department of Health and Environmental Sciences, wrote on March 19, 1990, that "Montana does not currently have a funded radon program and, therefore, has no public policy pertaining to radon."

By the early 1990s, interest at the state level was broader, but rarely intense. Every state had a contact person in charge of radon information. But the Conference of Radiation Control Program Directors cited only 15 states as having enacted or proposed legislation concerning radon. Almost all the legislation was limited to establishing certification requirements for testers or mitigators.[4]

Why the modest state activity? The attitude of many state health and environmental officials reflected the public's lack of concern. Moreover, radon studies and programs were not cost free. Although financial assistance was available from the federal government, states had to share expenses and divert resources from other projects. The absence of a sense of public urgency (despite the EPA's call for action), and the potential costs (despite availability of federal money), largely accounted for the restrained activity.

Contrary to the behavior of most, however, four states had undertaken vigorous action on radon before the 1988 act was passed: Florida, New Jersey, New York, and Pennsylvania. Despite severe budgetary pressures in the early 1990s, officials in these states continued to underscore the importance of their radon programs.

Florida

Florida was the first state in the nation to consider a statewide radon program. Areas of central Florida had long been known to contain phosphate ore, which has more uranium than most other soil and rock. News about radon in houses built on uranium tailings in the southwestern United States prompted concern in the 1970s about the situation in Florida. A survey in central Florida by the EPA and the Florida Department of Health and Rehabilitative Services showed homes with elevated levels of radiation. In 1979, the EPA administrator advised that remediation was necessary in some existing homes and that future construction should incorporate techniques to minimize radon concentrations.[5]

After the advisory, Florida's Governor Robert Graham appointed a task force to consider the matter. The task force proposed that a standard be established for acceptable indoor radon levels and that home occupancy be permitted only if the standard were met. In 1986, Florida became the first state to issue a regulation about human exposure to radioactivity from natural sources in the environment. The ruling held that annual average radon decay product concentration not exceed 0.02 working levels (equivalent to 4 picocuries of radon per liter of air) in new homes, schools, and commercial buildings.[6]

Public debate developed over whether the ruling should cover only reclaimed phosphate lands or the entire state. As a result, the state legislature mandated the Florida Institute of Phosphate Research "to direct a study of the entire state to identify all significant land areas of Florida where the rule should be applied." The study was completed in November 1987 and showed evidence of "elevated radon potential" in 32 of the state's 67 counties.[7] The governor then appointed a five-member peer review committee to assess the findings. The committee recommended that a radon notification statement be included in all property sales agreements, but that testing for radon be at the option of the buyer or seller.[8]

The recommendation was not unanimous, however. Gloria Rains, the lone environmental representative on the peer review committee, dissented. She urged that the action level be lower than 4 picocuries and that testing be required in all buildings before they could be sold. Several newspaper editorials applauded her stance, lamenting that she was "consistently outvoted by the special interests"—representatives of the real estate, building, and phosphate industries.[9]

In 1988, the state legislature enacted a statute that established 0.02 working level (4 picocuries) as the "standard" for existing housing—the same as the EPA's action level.[10]

Other recommendations of the peer review committee majority were also incorporated, including a requirement that information about radon be provided to parties in a property sale, but that testing be optional. A unique feature of the law was its creation of a radon trust fund to pay for radon programs. The money comes from a one-cent-per-square-foot surcharge on the construction or alteration of any building in the state.

As a result, Florida's radon appropriations have not been under the pressure that have affected radon budgets elsewhere. In 1991, the fund had amassed $3.6 million and, unlike in other states, radon activities continued apace.[11]

Pennsylvania

Indoor radon became a national issue after December 1984 when high levels were recorded in Stanley Watras' house in Boyertown, Pennsylvania. Michael Pyles, a field worker at the time for the state's Department of Environmental Resources (DER), helped with measurements in the Watras home. In 1992, having become chief of the radon division in the DER's Bureau of Radiation Protection, he recalled the frenzy caused by the unexpected findings.[12]

A survey of area homes by the EPA and DER after the Watras findings indicated that 45 percent had radon concentrations above 4 picocuries.[13] The affected dwellings were near the Reading Prong, a geological formation with higher than usual uranium deposits that extended through Pennsylvania, New York, and New Jersey. By the end of 1986, more than 18,000 Pennsylvania homes had been screened. Short-term measurements were taken in the basement, where concentrations are typically highest, and 59 percent showed radon levels in excess of 4 picocuries.[14]

Jason Gaertner, a community relations officer with the DER, presented a report in 1986 about the state's responses to the radon findings. They included acceleration of surveys to establish residential concentrations throughout the state, provision of public education materials, remedial action information and courses, and a state-backed low-interest loan program for mitigation in homes. The state had clearly embarked on an aggressive radon policy.

Toward the end of Gaertner's report, a single sentence appeared without elaboration: "The Pennsylvania Department of Health has conducted a limited lung cancer mortality survey in and around the Colebrookdale/Boyertown area [where homes with elevated radon levels were located] and found the rate of lung cancer deaths to be consistent with the rest of the state."[15] The comment appeared dissonant with the remainder of the report, which emphasized a broad front of action.

Gaertner concluded with the observation that the Pennsylvania

legislature had just passed a law to provide $1 million for a demonstra-
tion project to reduce residential radon levels. The project ended in
June 1988 after remediation efforts in 105 homes were completed.
Radon levels reportedly were reduced in almost all, but at a greater
cost than the $1,000 to $2,000 the EPA says it takes to mitigate a
home. According to the report, "remedial costs range from approximately $1,300 to $12,000, with an average of $3,700."[16]

The Pennsylvania annual budget for radon programs had climbed
from $1.5 million in 1985 to $3.5 million in 1986. By the end of the
decade it began to decline and in 1992 stood at about $700,000. (The
figure does not include several hundred thousand dollars provided by
the EPA for state projects.) Pyles, head of DER's radon division,
underscores the effect of the smaller budget. In the late 1980s, 21
state employees, including 10 field workers, were involved with testing, surveying, and other aspects of the state's radon program. In
1992, a half dozen were involved, including only two field workers. He
regrets the reduction in activity and attributes it to the weak economy
and to diminished interest in radon affairs by the "current state
administration."[17]

New York

In 1985, the New York Energy Research and Development Authority
began a statewide survey of several thousand homes. Although preliminary findings indicated that levels were not as high as reported in
Pennsylvania, in 1986 state officials announced the start of "several
new programs" concerning radon.

In addition to EPA projects in the state, a 1986 state law contained
provisions to make radon detectors available to homeowners "at reasonable cost," to investigate areas where elevated radon levels "might
be expected," to assess the impact of indoor radon levels on energy
conservation efforts, to establish a radon information hotline, and to
provide training in diagnosis and mitigation methods.[18] A statewide
survey based on short-term testing in basements was completed in
May 1990. One-third of the nearly 25,500 homes surveyed had concentrations of 4 picocuries or higher.[19]

Another statewide study offered a sharply different perspective,
however. Year-long testing in the basements of some 2,000 homes
showed 14 percent to have radon concentrations of 4 picocuries or
higher. Moreover, only 5 percent of the main living areas of these
homes had similar levels.[20] The findings were yet another confirmation of
the unreliability of short-term tests as indicators of annual averages.

Lawrence Keefe, a coauthor of the report on the short-term survey, remains convinced that radon is a widespread health threat. He was radon program coordinator for the New York Department of Health from 1986 to 1992, and expresses disappointment that the state's radon activities have been reduced: now some seven people in the Health Department work on radon compared to 14 a few years ago. He attributes the weakened activity only in part to budgetary decline. In large measure, he says, "the public has not been educated enough." But he is confident this will change. Keefe recalls that when people were first urged to stop smoking and to wear seat belts, few complied. "It took a long time for people to understand, and I think we'll see the same with radon." With radon expenditures in the state reduced from $2.5 million in the late 1980s to about $1 million in 1992, Keefe says, "We should keep working to educate, and eventually we'll get to where the radon problem is understood."[21]

New Jersey

Like the other two Reading-Prong states, New Jersey reacted rapidly to the news about the Watras home. In 1986, the state passed two radon laws that appropriated more than $4 million for programs. New Jersey's radon budget also has declined since then. In 1992, the state appropriated about $900,000 (apart from EPA money being spent for state projects). According to Tonalee Carlson Key, head of New Jersey's radon program development, interest in radon has not abated in the state. "There has always been a dedication here, as shown by continuing state appropriations for the program."[22]

The distinctive goals in the two 1986 New Jersey laws made clear an inconsistency that was blurred elsewhere. The first, enacted in January, provided $3.2 million to gain information. It called for studies to locate areas of radon concentrations, to try to develop remediation techniques, and to determine whether a relationship could be found between lung cancer and radon levels in New Jersey homes.[23] The mapping survey and remediation investigations were to be done by the state Department of Environmental Protection (DEP), and the epidemiological study by the Department of Health (DOH). (Late in 1991, the DEP became the Department of Environmental Protection and Energy [DEPE], but the former designation will be used throughout this chapter to maintain consistency.)

The studies had barely begun when in August a second radon bill became law. This one provided $1 million for the DEP to establish licensing procedures for firms that test and mitigate.[24] Its intention to protect consumers from incompetence and fraud seemed admirable.

But the provision implied that an abundance of knowledge existed about testing, remediation, and health risks in New Jersey homes. This was not the case, which was the reason for the first law—that more information about all these matters was needed. In May 1986, Donald A. Deieso, Assistant Commissioner of the DEP, acknowledged as much.

> [W]e do not adequately understand the scientific principles governing radon. Pathways into the home, geological factors, and remedial techniques remain in large part a mystery. Each day more is learned and each day more gaps in our knowledge are revealed.[25]

The law to establish certification standards for companies that test and remediate implied that knowledge about what to do was in place. Moreover, another state action reinforced this misperception. In December 1986, the New Jersey Housing and Mortgage Finance Agency announced a new low-interest loan program for homeowners with radon levels above 4 picocuries per liter of air. People who sought remediation could borrow up to $15,000 at 2 or 3 percentage points below private home-improvement rates. The agency set aside $7 million for the project and said more would be available as necessary.[26] (Paradoxically, difficulty in establishing certification standards delayed issuance of a regulatory code until January 1991.)

The loan program was announced 1 month after the DEP indicated it had developed an effective remediation technique. High radon levels had been discovered in March 1986 in the rural town of Clinton. During the summer and fall, the EPA and DEP placed an experimental piping and ventilation system in 10 homes. Radon readings fell sharply after the devices were installed, and DEP spokespersons called the department's work in Clinton a "success story."[27] Not all observers were enthusiastic, however.

John Spears, an architect who worked with government radon mitigation programs, was not especially impressed with the Clinton project. "If one fan didn't work, they would put in another, or some other mechanical devices until the numbers fell." "Look," he said without apparent cynicism, "if the EPA succeeds they continue to receive funding, so this project *had* to succeed." He questioned the long-term reliability of any system that depended on fans and motors.[28]

Even some ardent supporters of New Jersey's radon policies were cautious about encouraging homeowners to join in the program. Senator John Dorsey was the principal sponsor in the state senate of both 1986 radon bills. Yet early in 1987, while affirming his belief that radon tests could be performed accurately, he expressed less confidence about remediation. "The techniques dealing with high levels of radon are still being developed," he said. "Before making major expenditures I would think one would do well to wait."[29]

Thus, one of the authors of the law that would certify remediation firms suggested that people wait until techniques were better developed before using these companies. Testing, apparently, would have been only an information exercise. In recognizing the uncertainties about remediation technology at the time, Senator Dorsey inadvertently acknowledged the contradictions in his own bills. Yet New Jersey officials have spoken highly of the state's radon policies from the start. At a 1986 conference on "Radon and the Environment," Donald Deieso, the assistant commissioner for the DEP, referred to the state's "extremely dedicated professionals" who have "already made outstanding contributions in addressing [the radon] issue."[30] Senator Dorsey told the conference participants that "New Jersey has perhaps come up with the most responsive program of any of the other states."[31]

In 1986, after a survey suggested that one-third of homes in sections of the state had radon levels above 4 picocuries, the DEP advised all homeowners in northern New Jersey to test. A year later, when fewer than 5 percent had complied, the DEP urged again that everyone test—within the following 12 months.[32] The renewed plea had little more effect than the first. Gerald Nicholls, acting assistant director for radiation prevention programs, thought the public's reaction was not surprising. "It's the same reason that people don't stop smoking or don't lose weight." Their response to radon is understandable because "people just don't want to know bad news."[33]

The explanation that public apathy stemmed from a collective mental block, as Nicholls implied, may have been too facile. It ignored the contradictory nature of the state laws and uncertainties about testing and mitigation techniques. Moreover, it disregarded the effect of a highly publicized problem that the DEP was facing in the Montclair area.

Montclair and the Politics of Miscalculation

In December 1983, the DEP announced that several homes in the adjacent communities of Montclair, Glen Ridge, and West Orange contained unsafe levels of radon. The department believed the radon came from radioactive landfill over which the homes were built some 60 years earlier. The soil allegedly was transported there from the defunct U.S. Radium Corporation that stood in a nearby town.[34]

After estimating that 100 homes in the area had unacceptable levels of radon, a joint DEP-EPA task force considered approaches to remedy the situation.[35] With the thought that homes throughout the state might contain unsafe levels of radon from natural sources, the DEP intended to treat the Montclair area cleanup as a demonstration project. Removal of the soil would serve as a model for remediation elsewhere.

With the concurrence of the EPA, in February 1985 the DEP announced plans for the first step toward decontamination. The soil under and around 12 of the houses would be removed in the coming months at a cost of $8 million. After this demonstration phase, the remaining 80-odd homes would be cleaned up in cooperation with the EPA at a cost not yet stipulated. Thomas A. Pluta, the DEP's designated director of the effort, said that while the soil was being removed, the occupants would live in a hotel for 2 to 4 weeks at the state's expense.[36]

Where to Ship the Soil—1985

Initially, the DEP intended to store the dug-up soil in a West Orange armory. But after objections from local residents, it sought an out-of-state solution. By the time the plan was announced in February, the department ostensibly had an understanding with the operators of a waste facility in Beatty, Nevada, to receive the soil.[37] The effort would involve placing the dirt in 55-gallon steel drums, shipping the drums 8 miles to a lot in Kearny, and from there by rail to Nevada.

Governor Thomas H. Kean issued an executive order to override a Kearny zoning law that would have prohibited the placement of the radioactive soil in that town. Residents organized a petition drive and held a march to protest the governor's action. But the DEP insisted that the drums would sit in Kearny for only a brief period, and overt opposition subsided. Michael Beard, Kearny's environmental officer, helped assuage local concerns. "I'm sure [the DEP] is doing it the best way possible," he said, "and that the health threat is at the minimum."[38] These may have been the last approving remarks offered by anyone about the DEP's plan.

Although the department thought it had an understanding with Nevada authorities, it learned in July after excavation had begun that local and county officials in Nevada were seeking a federal court order to prohibit the shipment. Nevada's Governor Richard H. Bryan joined them, declaring that Nevada was "not going to be the nuclear dumping ground for the country."[39] In October, the U.S. Supreme Court overruled a federal court decision in favor of New Jersey's claim and denied permission to begin shipment. Federal Energy Secretary John S. Herrington then announced that New Jersey should find a dump for the soil within its borders.[40] New Jersey officials decided to pursue legal efforts to force Nevada to accept the radioactive dirt. But they had no idea if or when they would be successful. For the moment the soil disposal plan was stymied.

Meanwhile, excavation around eight of the affected houses had been completed at a cost of $600,000 each. Although the rest of the project was suspended, almost 15,000 barrels had been filled with radioactive soil. Nearly 10,000 were shipped to Kearny, and the remaining 5,000 were stacked on the lawns of four of the homes in Montclair. At the end of 1985, occupants of the four homes were still

Some of the 5,000 barrels of radium-contaminated soil, covered by tarpaulins, outside houses on Franklin Street in Montclair, NJ.

in the temporary quarters they had occupied since August. No one knew when they would be permitted to return.

Around this time, the DEP had raised its estimate of radon-contaminated homes in the area from 100 to 200. The year ended badly for the residents in the Montclair area and for the department's soil removal program. The next year would be no better.

Where to Ship the Soil—1986

In January, the DEP proposed moving the radioactive soil to Picatinny Arsenal in New Jersey. There it would be mixed with clean soil to lower the radioactive concentration. The Department of Defense rejected the idea, however, saying its facilities could not be used for nonmilitary purposes.[41] At the same time, the DEP was facing the added pressure of a lawsuit by Montclair to force removal of the barrels from the lawns.

In June, the department announced that the soil would be moved to a quarry in Vernon Township where it would be blended with uncontaminated dirt. Vernon residents and township officials, who had not been consulted before the announcement, were furious. Thousands attended a rally and vowed to block the shipment.

The DEP's actions, especially its unwillingness to discuss the matter with local representatives in advance, drew anger from across the state. Six bills were introduced in the state assembly to block the DEP's plan to move the soil to Vernon Township. Alan J. Karcher, the state assembly's Democratic minority leader, unceremoniously expressed the sentiment of legislators from both parties. "This is probably the most high-handed situation that the DEP has ever done.... The DEP has an arrogance of power. They don't give a damn about the people."[42]

The DEP continued to insist that the site plan in Vernon would result in virtually no greater level of radioactivity than normally found in dirt. Its contention may have been scientifically sound, but the department's approach had doomed the plan. Its failure to educate the affected public or to consult with local officials in advance—to build a consensus of support—invited the antagonistic reaction.

Vernon residents display dissatisfaction with state's plan to ship radioactive barrels to their township.

AL PAGLIONE/THE RECORD

In November, the DEP retreated. The department's commissioner, Richard T. Dewling, said the plan to move the soil to Vernon would be abandoned, and he promised "a new process involving extensive public input" for future proposals.[43] As another year ended, the 15,000-odd barrels remained in Kearny and Montclair. The displaced Montclair families were still unsure when they could return to their homes. They had become bitter and depressed, as were others in the neighborhood who continued to live in the midst of ostensibly unsafe radon concentrations.[44]

Commissioner Dewling acted on his pledge by appointing a Radon/Radium Advisory Board (RRAB) in December 1986. The 10-member board, which included representation from communities, environmental groups, and state agencies, was asked to suggest a disposal site for the soil. The RRAB then set an agenda to hear expert testimony and comments from the public at a series of meetings, after which it would issue a report.

Where to Ship the Soil—1987

The RRAB had barely begun its deliberations when in February 1987 a state superior court judge created another complication for the DEP. Sitting in Essex County, the county in which Montclair lies, the judge ruled that the DEP had to remove the drums from the four lawns by May 15.[45] The DEP decided not to appeal and said the soil would be moved to another temporary site before the deadline. Montclair residents, having been disappointed by unfulfilled past promises, reacted skeptically. Samuel Pinkard, the head of Montclair's Radon Task Force (a body established 3 years earlier by the town council), declared, "At the rate the DEP does things, they could remove one barrel and take who knows how many years to remove the rest."[46]

Members of the newly established RRAB were also upset by the DEP's declaration. Several felt their efforts would be impaired because "there is bound to be a large political outcry when the DEP selects an interim storage site," as one member said. Once the DEP tries again to impose a solution on a community, "it will hurt our credibility and endanger our efforts to find a long-term volunteer storage solution."[47]

Criticism from the board created by the DEP to mute criticism was unwelcome news for the department, but there was more. The EPA had just issued a report warning that the number of homes contaminated by radon in the Montclair–Glen Ridge–West Orange cluster had been underestimated. Instead of 200 homes, the figure was raised to 450 or more. The EPA's field representative, John

Frisco, said the agency had not yet determined the new findings' "outer limits."[48]

As the court-imposed May 15 deadline approached, tension among Montclair's residents was palpable. With 1 week to go, a rumor surfaced that the DEP would move the barrels from the lawns to a nearby public park. In response to inquiries, the DEP refused to rule out the possibility, and 750 people held a protest rally at the park. The rumored site was across the street from an elementary school, and Montclair's superintendent of schools vowed to close the school if the barrels were placed there.[49] Two days before the deadline, the community was in an uproar, and *The Record*, northern New Jersey's largest circulation newspaper, chastised the DEP and Governor Kean for remaining silent. Like many Montclair residents, the newspaper thought that moving the radioactive soil to the park would be "sheer lunacy."[50]

At the last moment the court extended the deadline by 1 month. The park site idea faded away as the DEP revealed it had yet another plan. Drawing from suggestions by its Radon/Radium Advisory Board, the department said it was approaching several communities with an offer of $6 million to store Montclair's 5,000 drums temporarily. The only town identified was Kearny, which was still hosting nearly 10,000 barrels from the original batch. In response to the news, the Kearny municipal council not only unanimously rejected the offer, but authorized a suit to force the DEP to remove the 10,000-odd barrels from its land. The community's indignation was expressed by Councilwoman Jean A. Byrnes, who noted that the town had "suffered so many environmental insults that to accept more [would be] a travesty."[51]

Immediately afterward, the DEP said that it was simultaneously pursuing another possibility. Despite the Pentagon's rebuff 2 years earlier when asked that Picatinny Arsenal receive the soil, Commissioner Dewling was again seeking help from the Defense Department. He now requested space at any of the six military facilities in the state.[52]

By the beginning of June, no town seemed interested in taking the drums for the DEP's money offer, and no further word appeared about storage at a military base. Then on June 4, Commissioner Dewling announced that the Montclair soil would be moved temporarily to a state-owned wildlife area in the Pinelands. The site is in Ocean County near Jackson Township. The DEP said it would start shipping the drums days before the court-ordered June 15 deadline. No sooner was the plan announced than environmentalists, hunters, and local officials sought an injunction to stop the DEP from acting.[53]

Once again the department had announced its decision without warning, without preparing local and other interested parties, a formula that had invariably produced failure in the past. Predictably, Ocean County officials and residents of Jackson Township were enraged. As hundreds demonstrated, Jackson's Mayor Daniel J. Black insisted that the community would not be "dumped on."[54] On June 6, a state superior court judge in Ocean County barred the DEP from bringing the soil to the wildlife preserve pending a hearing. The judge in Essex County then lifted his June 15th deadline and agreed to have the case consolidated with that in Ocean County.[55]

Hopes for a solution seemed as distant as ever. Secretary of Interior Donald P. Hodel informed Governor Kean that he opposed the movement of the radioactive material to the wildlife refuge. This would violate the federal-state Pinelands National Reserve management agreement, the secretary said.[56] The governor disagreed, and the DEP reaffirmed its intention to move the soil to the Pinelands refuge. But by the end of June the state legislature sought to foreclose the possibility. A large majority in each house voted for a bill to prohibit the DEP from storing radium-contaminated soil in any state wildlife refuge or on any site in the Pinelands.[57] Although the governor vetoed the measure, any effort to move the material to those locations would have fueled even more disruption.

Thus in July 1987, the DEP was still without an imminent option. It had just reopened its Office of Public Participation, which had been inactive for years, but this was of little help with the immediate problem. The Radon/Radium Advisory Board continued methodically to hold hearings and deliberations. Although the RRAB had produced a draft paper on criteria for temporary storage of the soil, it did not plan to issue final recommendations for months.

The sense of urgency and frustration in many communities was growing. In response, Governor Kean appointed Richard G. Sullivan as a special consultant to search for "any reasonable alternatives" for storing the drums. Sullivan had been the state's first environmental commissioner, from 1970 to 1974.[58] But whether any additional person or group could improve the process of site selection seemed doubtful.

Sullivan's appointment could prove useful only if it enhanced popular acceptance of the ultimate choice. The principal issue was no longer a matter of site selection, but whether the DEP's credibility had so withered that *any* in-state solution it proposed could be carried out. The DEP erred about popular reaction so often that skepticism now shadowed its scientific and technical calculations as well. This was especially true in communities involved with the soil problem, either as source (Montclair, Glen Ridge, West Orange) or potential receiver (Kearny, Vernon Township, Jackson Township). Samuel Pinkard had become convinced that DEP officials "don't tell the truth."[59]

In September 1987, the DEP seemed at last to have found a way to rid Montclair of its 5,000 drums. In a bizarre arrangement, the department contracted with a company for $4 million to ship the soil to Oak Ridge, Tennessee. There it would be mixed with highly radioactive materials. The mix could then be sent to a federally approved depository in Richland, Washington, that accepts only high-level waste.[60]

The expensive and circuitous arrangement drew ridicule (*see illustration below*), but the citizens of Montclair were relieved to see a few hundred barrels shipped out during the following weeks. By mid-1988, 3 years after excavation began, all the barrels had been removed from Montclair and Kearny.

This cartoon, which appeared originally in *The Knoxville* (Tenn.) *News-Sentinel*, was reprinted in *The Record* (Hackensack, NJ) on August 9, 1987.

But the story did not end there. Members of Montclair's Radon Task Force thought more dirt should be removed. DEP officials disagreed, saying that they were satisfied the job was complete and that it was enough to fill the existing hole with clean soil. In June 1989, the EPA tried to break the impasse. It announced a $53 million plan to dig up and haul away more soil from around and under 26 houses in the area, including the four empty ones, and to install mitigation devices in 140 others. The EPA said it would complete the work in 3 years.[61]

The following year, the agency expanded its plan and promised to remove the surrounding soil from 400 houses. The revised effort would take 10 years, according to federal officials, and would cost at least $250 million.[62] The New Jersey DEP would not be a participant in the cleanup program.

By 1991, the EPA had removed the contaminated dirt from under 15 houses, including the four that had been empty since 1985, and shipped it to a waste depository in Utah. But years of neglect had made the four homes uninhabitable. One was rebuilt and its owners were able to return. The other three were torn down after the EPA purchased them from their displaced owners. The families found permanent lodging elsewhere, and their former locations remain empty lots. Removal of the soil from another 50 houses was scheduled to begin in late 1992. Samuel Pinkard, Montclair's Radon Task Force chairman, expressed satisfaction with the EPA's accomplishments, and relief that the DEP was no longer involved.[63]

Assigning Blame

The DEP seemed as unwilling to accept blame for the Montclair radon problem as the affected communities were willing to assign it. The department chose to denigrate the attitudes of the state's residents while publicly ignoring its own miscalculations. Commissioner Dewling, frustrated by the unrelenting soil problem, accused politicians and the press of stoking a "not-in-my-backyard" mentality among New Jersey residents. He said this attitude was "the single major cause of environmental gridlock."[64]

Similarly, the department's assistant commissioner, Donald Deieso, attacked unnamed scientists and activists for encouraging emotional reactions among the public. "Cloaked in the mantle of environmentalism," he said, "they cater to the fears of a community and play on risks and chemophobia. In short, our residents are being manipulated...."[65] As to the plight of the four families displaced from their homes, James Staples, a DEP spokesman, said, "I think they have been dealt a bad hand by fate, not by us."[66]

Exaggerated fears among citizens doubtless helped undermine what may have been scientifically sensible plans. But popular rejection was largely primed by the DEP's manner of informing. The DEP's favored technique was the surprise announcement. The department's record of failure to consult and educate in local communities *before* announcing its policies was dedicated and consistent. An abrupt declaration by the DEP about where the soil would go was invariably followed by local demonstrations in opposition, lawsuits, and eventual revocation of the plan. The repeated sequence prompted Samuel Pinkard, chairman of Montclair's Task Force on Radon, to muse about the state's environmental leadership and how "any person with mild intelligence [can] make the same mistake four times."[67]

Barbara Haver began her 4-year term as president of Montclair's League of Women Voters in 1983, the year the radon issue surfaced in the town. From the start, the issue was an important part of her organization's agenda. At numerous town meetings, crowded with local residents, she heard a succession of DEP officials speak of their plans. "By and large," she recalls, "townspeople felt the DEP members were neither knowledgeable nor sensitive."[68]

Rosemary Pusateri expressed similar sentiments. From 1988 through 1991, she chaired Montclair's Environment Advisory Committee, a body appointed by the town council. A member of the committee since 1984, she also served on the town's Radon Task Force. Among the DEP officials she met, some impressed her as caring individuals, but "regardless of their intentions, most seemed inept and callous." There was townwide concern during this period, she says, that "the DEP seemed in over its head."[69]

Seeking Trust

Neil Weinstein and Peter Sandman offer a more sanguine view of New Jerseyites' attitudes about the DEP. Weinstein, a professor of psychology at Rutgers University, and Sandman, a professor of environmental journalism there, conducted inquiries in the 1980s under contract from the DEP about the public's response to radon risk.[70] The final report of their 1988-1989 survey, coauthored with Nancy Roberts, a human ecologist, indicated that most respondents "underestimated radon risks," that their "emotional responses to radon were mild," and that they rated the DEP as "the most accurate source of radon information."[71]

The positive inference about the DEP is at odds with the views cited here of citizens in communities involved with the Montclair soil issue. The Weinstein-Sandman-Roberts findings seem explainable in part because none of their questionnaire items contained references to the Montclair issue.

Moreover, residents in communities involved with the Montclair soil problem were thinly represented in the survey. The authors chose respondents disproportionately from Tier 1 areas. (The DEP classifies Tier 1 locations as those most likely to have elevated radon concentrations from natural sources, Tier 2 less likely, and Tier 3 least likely.) Among the several communities involved with the Montclair soil issue, only one, Vernon Township, was in Tier 1. A study that included a larger representation of residents from the affected areas likely would not have yielded the same results.

State officials prefer to separate the Montclair problem from other radon issues. They distinguish between radon in the Montclair area homes, apparently caused by man-made sources, and accumulations caused by natural sources. In terms of mitigation techniques and goals, the warrant for distinction is not clear. But whatever its validity, the Montclair experience is unarguably a consequence of a radon mitigation program not carefully thought out. It is a reminder of the need for circumspection; that well-intended policies may lead to terrible consequences. It is also instructive about the harm that can come from concealing information. The DEP thought it knew what was best for the citizens and often acted without informing or consulting them. In the end, this engendered only distrust.

Paradoxically, the most striking recommendation that Weinstein, Sandman, and Roberts made in their 1989 report was that the DEP should refrain from informing the public about uncertainties concerning radon. To avoid giving people excuses not to act, the authors suggested that the DEP "put forward an unambiguous 'united front.'" About what the front should be there was no doubt.

> *The DEP should try to avoid public perception that radon risks or radon mitigation techniques are debatable or uncertain* (Emphasis in original).[72]

Sandman and Weinstein have also been consultants on risk communication to the EPA, and their advice was echoed in the principal message at the EPA-sponsored radon symposium in 1991. There the keynote speaker admonished his audience "to speak with one voice to the public."[73]

Notes

1 Indoor Radon Abatement, Amendment of the Toxic Substances Control Act, Title III, Pub. L. No. 100-551 (S. 744), §305(a), Oct. 28, 1988.

2 Ibid., §§305 (b), 306 (j).

3 Joseph E. Rizzuto, "New York State Energy Research and Development Authority Radon Program," and Jason Gaertner, "Commonwealth of Pennsylvania Radon Monitoring Program," in *Radon and the Environment*, eds., William J. Makofske and Michael R. Edelstein, Conference Proceedings, Ramapo College of New Jersey, May 8–10, 1986 (Mahwah, NJ: Institute for Environmental Studies, Ramapo College, 1987).

4 State Radon Legislation, Jan. 1992, (mimeo), accompanying *Radon Bulletin*, published by the Conference of Radiation Control Program Directors, Inc., in cooperation with the U.S. Environmental Protection Agency, Vol. 2, No. 2 (Winter 1991).

5 Gordon D. Nifong, "Perspective," *Florida Statewide Radiation Study*, prepared by Geomet Technologies, Inc., under a grant sponsored by the Florida Institute of Phosphate Research, March 1989, iii.

6 Ibid., ch. 2, p. 3.

7 Ibid., iv, and ch. 7, p. 1.

8 Ibid., v.

9 "Will Florida Cut Radon Risks?" *St. Petersburg Times*, 10 Dec. 1987, 26A; similarly, "Unnecessary and Unacceptable," *Sarasota Herald-Tribune*, 2 Nov. 1987, 10A.

10 CS/CS/HB 1420, 1st Engrossed, amending Fla Stat. §404.056, creating Fla. Stat. §553.98.

11 Interview, Norman Michael Gilley, Public Health Physicist Manager, Florida Dept. of Health, Apr. 16, 1992.

12 Interview, Apr. 16, 1992.

13 Jason Gaertner, "Commonwealth of Pennsylvania Radon Monitoring Program," in Makofske and Edelstein, 367.

14 Thomas M. Gerusky, "The Pennsylvania Radon Story," *Journal of Environmental Health*, Vol. 49, No. 4 (Jan.–Feb. 1987), 197–99.

15 Gaertner, in Makofske and Edelstein, 372.

16 Pennsylvania Department of Environmental Resources, Bureau of Radiation Protection, *Final Report of the Pennsylvania Radon Research and Demonstration Project* (Harrisburg, PA: Pennsylvania Dept. of Environmental Resources, June 1988), ch. 1, pp. 1–3.

17 Interview, Apr. 16, 1992; Michael R. Edelstein, "The State Radon Survey," in Makofske and Edelstein, 331.

18 William Condon, "New York State Department of Health Radon Program," in Makofske and Edelstein, 380–81.

19 Charles Laymon, Charles Kunz, and Lawrence Keefe, *Indoor Radon in New York State: Distribution, Sources and Controls*, Technical Report (Albany, NY: State of New York Dept. of Health, Nov. 1990), 42.

20 R.L. Perritt, T.D. Hartwell, L.S. Sheldon, B.G. Cox, C.A. Clayton, S.M. Jones, M.L. Smith, and J.E. Rizzuto, "Radon-222 Levels in New York State Homes," *Health Physics*, Vol. 58, No. 2 (Feb. 1990), 147–55.

21 Interview, Apr. 20, 1992.

22 Interview, Apr. 21, 1992.

23 N.J. Pub. L. No. 1985, Ch. 408 (Jan. 10, 1986).

24 N.J. Pub. L. No. 1986, Ch. 83 (Aug. 14, 1986).

25 Donald A. Deieso, "An Overview of the Radon Issue in New Jersey," in Makofske and Edelstein, 414.

26 *Record* (Hackensack, NJ), 24 Dec. 1986, C-1.

27 On November 17, 1986, the DEP and EPA held a press conference at which an information packet was distributed under the heading "Clinton Success Story."

28 Interview, Nov. 19, 1986, and presentation at a seminar on "Radon: Its Impact on You and Your Municipality," sponsored by the New Jersey Department of Environmental Protection and the American Association of Radon Scientists and Technologists, Atlantic City, NJ, Nov. 19, 1986.

29 Interview, Jan. 3, 1987.

30 Deieso, in Makofske and Edelstein, 414.

31 Senator John Dorsey, "The New Jersey Radon Program: A Legislative Perspective," in ibid., 410.

32 *Record*, 11 Sept. 1987, A-1.

33 Interview, Nov. 19, 1986.

34 U.S. Radium employees painted radium on watch dials. John Dughi, a spokesman for the company's successor, the Safety Light Corporation, rejected the DEP's allegation. He claimed that there was no connection between the radioactive soil and the previous radium company, and that no radium left the plant site after its closing in the 1920s. *New York Times*, 17 Feb. 1985, 57. An investigation by a Rutgers University geologist undertaken for a lawyer suing Safety Light concluded that the soil came from the U.S. Radium Corporation.

Richard K. Olsson, "Geologic Analysis of and Source of the Radon Contamination at the Montclair, West Orange, and Glen Ridge Radium Contaminated Sites," Final Report for Wayne D. Greenstone, attorney at law, Newark, NJ, June 19, 1986.

35 "Siting Criteria for Temporary Storage of Containerized Radium-Tainted Soils," Draft of a Report by the Radium/Radon Advisory Board to the New Jersey Department of Environmental Protection, June 8, 1987, 10 (mimeo).

36 *New York Times*, 17 Feb. 1985, 57.

37 "Siting Criteria for Temporary Storage of Containerized Radium-Tainted Soils," 10.

38 *New York Times*, 30 June 1985, NJ-15.

39 Ibid., 10 Aug. 1985, 27.

40 Ibid., 6 Dec. 1985, B-7.

41 Ibid., 30 March 1986, NJ-18.

42 *Record*, 14 Sept. 1986, A-23.

43 *New York Times*, 27 Nov. 1986, B-1.

44 Ibid., 21 Aug. 1986, B-2; *Record*, 7 Dec. 1986, A-1.

45 Judge Murray G. Simon's ruling in response to the Montclair suit was reported in *Record*, 8 Feb. 1987, A-29.

46 *Record*, 23 March 1987, C-20.

47 Statement by John Lehman, *Star-Ledger* (Newark, NJ), 25 March 1987, 17.

48 *New York Times*, 26 March 1987, B-1; *Record*, 26 March 1987, B-2.

49 *Record*, 7 May 1987, B-1.

50 Ibid., Editorial, 13 May 1987, A-22.

51 Ibid., 27 May 1987, A-1, and 28 May 1987, B-1.

52 Ibid., 28 May 1987, B-1.

53 Ibid., 5 June 1987, A-14; *New York Times*, 5 June 1987, B-2.

54 *New York Times*, 6 June 1987, A-29.

55 *Record*, 10 June 1987, A-9.

56 *New York Times*, 11 June 1987, B-4.

57 The Senate vote was 26-7, and the Assembly's was 42-17. *Sandpaper* (Ocean County, NJ), 24 June 1987, 22.

58 *Record*, 16 June 1987, A-3.

59 Interview, Dec. 21, 1986.

60 *Record*, 11 Sept. 1987, A-3.

61 Robert Hanley, "U.S. Announces Plan to Remove Radioactive Soil," *New York Times*, 1 July 1989, 26.

62 Anthony DePalma, "U.S. to Clean Radium in Three Towns," *New York Times*, 7 June 1990, B-1.

63 Interview, Apr. 26, 1992.

64 Governor Kean agreed with his commissioner. *Record*, 14 June 1987, A-8.

65 *New York Times*, 5 July 1987, A-8.

66 Caryl R. Lucas, "Owners of Radon Houses Still 'Out in the Cold,'" *Star-Ledger*, 28 Feb. 1989, 28.

67 *Record*, 14 June 1987, A-8.

68 Interview, Apr. 26, 1992.

69 Interview, Apr. 27, 1992.

70 Neil D. Weinstein, Peter M. Sandman, and M.L. Klotz, "Public Response to the Risk from Radon, 1986," Research Contract C29543 (Division of Environmental Quality, New Jersey Dept. of Environmental Protection, Jan. 1987); Neil D. Weinstein, Peter M. Sandman, and Nancy E. Roberts, "Public Response to the Risk from Radon, 1988–1989," Research Contract C29418 (Division of Environmental Quality, New Jersey Dept. of Environmental Protection, Nov. 1989).

71 "Public Response to the Risk from Radon, 1988–1989," ch. 2, pp. 1–2.

72 Ibid., ch. 2, p. 18.

73 John R. Garrison, Managing Director, American Lung Association, Keynote Address, Environmental Protection Agency, *The 1991 International Radon Symposium on Radon and Radon Reduction Technology*, Philadelphia, PA, Apr. 2, 1991.

7

Private and Public Interests

IT IS AXIOMATIC THAT POLITICS IS BUILT ON INTERESTS. "One of the fundamental truths about politics," Karl Deutsch wrote, "is that much of it occurs through the pursuit of *interests* (original emphasis) of particular individuals or groups."[1] Equally unassailable is the generalization by Anne Hiskes and Richard Hiskes that "the job of democratic institutions is to ensure that the interests of all people are justly considered."[2] But the challenge in a democratic society is to translate the various interests into public policy.

The public interest is not monolithic. Individuals and groups commonly evoke their own interests in the guise of promoting the public good. Transparent as the exercise sometimes is, it is a necessary ingredient of democratic politics. One measure of a democratic system is the degree to which competing interests may become organized. A second is the means by which they receive a hearing. In the United States, both tasks are accomplished largely through voluntary associations, which Gabriel Almond and Sidney Verba described as "the prime means by which the function of mediating between the individual and the state is performed."[3]

Interest groups try to influence policy by a variety of methods. They may disseminate information to the public and to government officials, make financial contributions to political campaigns, hold rallies and marches, or otherwise demonstrate to call attention to their cause. In seeking to affect radon policy, groups have expressed their positions largely by approaches to elected and appointed government officials.

In assessing the roles of groups and agencies interested in radon policy, this chapter reviews testimony by disparate groups at the nine congressional hearings held between 1985 and 1992. Although a range of views was offered, most witnesses reflected the belief that indoor radon constitutes a serious health hazard throughout the United States. The positions of the groups are examined from the perspective of their apparent economic interest in the matter. Three categories are evident: financial gainers, financial losers, and nonprofit advocacy associations.

Financial Gainers

As might be expected, groups that profit financially from radon management have supported an aggressive policy approach. The most obvious beneficiaries are companies that provide testing and remediation services, but others stand to gain as well. Insurance companies profit from writing liability policies for radon remediators as well as for sellers and agents involved in real estate transactions. Lawyers gain from writing contracts and representing litigants over radon issues.

Although insurance and legal groups did not offer formal testimony at the congressional hearings, the prospect of their financial gain was recognized soon after radon became a national issue. This was manifest, for example, at a 1986 forum cosponsored by the New Jersey Department of Environmental Protection (DEP) and the American Association of Radon Scientists and Technologists (AARST), an organization representing radon management firms. The subject was "Radon: Its Impact on You and Your Municipality," and the thrust was to encourage radon management activities. When a speaker predicted that radon management "will be a full-employment industry for lawyers [and others] in the audience," his remark drew smiles and nods of agreement from the 80 people in attendance.[4] Commercial interests that stood to gain directly have continued to support aggressive radon policies.

Niren Nagda of Geomet Technologies in Maryland typified the views of spokespersons for companies that monitor and mitigate indoor radon levels. At a congressional hearing in 1987, he testified in support of the bill to amend the Toxic Substances Control Act "to assist states in responding to the threat to human health posed by exposure to radon." But he emphasized the need for a more expansive and costly program than contained in the bill.[5]

Nagda wished the bill had provided aid "even to states where...the seriousness and extent of radon exposure may be unknown." Moreover, although the EPA and the Surgeon General had not yet advised comprehensive national testing, he urged that the bill's proposed study of radon in schools be expanded to include every county in every state. Finally, he recommended that the appropriation for the school survey be raised from $500,000 to $1 million.[6]

Another spokesman for the radon industry, Marvin Goldstein, informed members of Congress that the country was facing a crisis because of radon. At the Senate hearing in 1990, he offered a written statement on behalf of AARST that radon "is by far the largest indoor environmental health risk that this nation faces today."[7] The following year, he told members of the House that "there is no question that radon is the most critical indoor carcinogen to be dealt with in this century."[8]

Goldstein dismissed concerns that testing devices could be irresponsibly manipulated or that short-term tests might not represent year-long concentrations. His organization took "great pride," he said, in its ability to discover tampering with radon tests. Moreover, "these are tests which can be completed and results provided by professional testers within 5 days. It doesn't take weeks."[9] He offered no evidence to support the implied validity of short-term testing, nor did any congressman ask him to justify his claim.

After warning again that radon "is the largest single indoor environmental health danger known to exist," Goldstein invoked dramatic imagery to make the point. He noted in his 1990 testimony that the 1988 Stewart McKinney Homeless Assistance Act required the Department of Housing and Urban Development (HUD) to develop a radon policy for the 3 million homes covered by the act. But because HUD had not yet complied with the mandate, "almost three people a day are dying of radon-induced lung cancer."

Goldstein ended his testimony with a plea that overstated the Environmental Protection Agency's presumption that radon accounts for 5,000 to 20,000 annual lung cancer deaths. "Since we know that radon kills over 20,000 Americans each year, let's protect our families and ourselves by testing for radon and mitigating any radon problem found."[10]

At a Senate hearing in 1991, James W. Krueger testified on behalf of AARST that the health risk to the American public posed by radon is "grave." He said that health and environmental agencies "concur that radon is the most critical indoor carcinogen to be dealt with in this century." He urged Congress to "seize the moment" and enact legislation that would require testing and remediation in all buildings funded by federally insured mortgages.[11]

Krueger also castigated HUD for not complying with the radon testing provisions of the McKinney Homeless Assistance Act. He estimated that about 12 million people lived in the 3 million McKinney Act homes. "Using EPA's own figures," he said, "that translates into about 800 deaths per year." He then raised the specter of lawsuits against the federal government for its failure to act.

> Clearly HUD has not responded in compliance with the congressional mandate. There is a serious potential liability because the Federal Government knows the magnitude of the risk and its own responsible agency has not responded to its mandate. As many as two or three people a day may be dying of radon-related lung cancer in McKinney Act housing. For every day HUD delays, its potential responsibility grows.[12]

Krueger's written submission to the subcommittee concluded with a list of proposals that AARST wished to see become law. In addition to mandatory testing and mitigation for all federally funded housing, the list urged that EPA be required to enact a "radon testers certification program" and to "define maximum permissible levels of radon in indoor air and in drinking water."[13]

Also in 1991, Keith S. Fimian, another radon industry spokesman, criticized the EPA's suggestion that an "inexpensive" do-it-yourself test is as effective as one performed by a company. "This of course is not true," he said in a written statement. Fimian implied that radon companies usually use electronic instruments rather than the passive charcoal devices in the self-test kits. Though offering no evidence, he said that unlike the homeowner's technique, the companies use "greatly enhanced technology, which is to say accuracy, reliability, and quality assurance; in short, a better test."[14]

At the 1992 hearing, Richard D. Martin of DMA-Radtech, a Pennsylvania radon testing firm, testified along the lines that industry spokespersons had at other hearings. He urged that EPA be given more legislated power to address "the number one environmental health threat facing the Nation today."[15] His testimony, however, which was offered on behalf of AARST, contained exaggerations and errors.

In bold print, Martin's written statement said: "Evidence exists that children are 300 percent more susceptible to the influence of radon than are adults."[16] In fact, a National Academy of Sciences study concluded that an "assumption of either an enhanced or a reduced effect for [radon] exposure during childhood is subject to substantial uncertainty."[17]

Martin further claimed that compared to other countries, "the United States lags behind in developing an ambitious program to deal with the radon issue head on." As an example, he noted that in

Sweden "53 percent of existing homes with elevated radon levels have been remediated." He cited a 1990 article by Gun Astri Swedjemark as his source of information.[18]

In fact, Swedjemark (and her coauthor Astrid Mäkitalo) were referring to a sample of 1,463 Swedish homes with radon levels above 20 picocuries per liter of air. Mitigation reduced radon concentrations in 53 percent of them to levels below 10 picocuries. (In only 12 percent of those homes were levels reduced below 4 picocuries.[19]) Contrary to Martin's inference, the article said nothing about Sweden's 3.8 million homes, very few of which have been tested for radon, let alone remediated (see Chapter 10).

Thus, spokespersons for the radon testers and fixers emphasized that indoor radon presented a serious problem that required their unique services. Yet their testimony often exaggerated the ostensible problem, presented speculation as fact, and was sometimes plainly wrong.

Financial Losers

Organized groups that might suffer financial losses as a result of aggressive radon policies tend to be associated with the real estate industry. They largely represent home builders and realtors. Home builders face additional expense as they alter construction to accommodate new radon regulations. (Presumably they could pass the costs to home purchasers by raising prices, but this was not discussed at the congressional hearings.) Moreover, as with realtors, they would suffer financial loss if a transaction were delayed because of testing results or mitigation.

One expects, therefore, that real estate interests would be unenthusiastic about a strong radon policy. This was generally the case. But while expressing concern about economic consequences, they invariably invoked the "good citizen" image of caring about peoples' health and safety.

Realtors worried about the effects a vigorous radon policy might have on their business when the subject first drew national attention. At a congressional hearing in 1985, Richard Barnola, a Pennsylvania real estate agent, spoke of his mixed emotions. "I am not only concerned about the effects of radon on the health of my family and myself and everybody else, but also with its economic impact." His territory included the Reading Prong area, the geological formation with elevated radon concentrations.[20]

Publicity about the Reading Prong's supposed danger, according to Barnola, had "created a great degree of anxiety, fear, and confusion among property owners, not to mention the fact that I am sure it has scared the hell out of potential property buyers." Frustrated by the uncertainty, he questioned whether living in the areas was truly hazardous. "Based on what I have been able to determine, we don't really know enough right now about the incidence, the measurement, the effect, and the geography of radon" to make that assumption.[21]

Richard Tracy of Ryan Homes, a home construction company headquartered in Pennsylvania, demonstrated similar concerns at a hearing in 1987. He noted that many people remained apathetic about residential radon, but he despaired that others had become panicky. "In our business, that panic touches both the builder (or seller) and the buyer, and it is often caused by inflated costs quoted for remedial work. There have been situations where unscrupulous contractors charge thousands of dollars for work that is either not necessary or worth only a few hundred dollars."[22]

He testified that his company's construction practices followed EPA recommendations, and this enabled customers later to install piping and fans to vent radon if they wish (at their own expense). He was cautious about suggesting policy direction and supported additional radon research rather than a crash action program.[23]

Barry Rosengarten testified in May 1988 and again in May 1990. Representing the National Association of Home Builders (NAHB), his comments at the earlier hearing were more conciliatory than at the later one. The flavor of his organization's position in 1988 was typified by the statement that "[w]e do not wish to become involved in prolonged public debates about the health effects of radon. NAHB builders simply want to build the best buildings we can so that the public can have safe, healthy, and affordable housing."[24] His concern about the health-affordability balance seemed unexceptional and bland.

At the same time, he wove into his testimony several uncertainties about the state of knowledge concerning radon. He suggested that "current testing technology does not allow future indoor radon levels to be predicted from a site test." He noted that "much confusion exists regarding the meaning of short-term (3–7 days) home test results. Radon levels can fluctuate on a seasonal, daily, or even hourly basis." He concluded that "because of the unreliability of short-term test results... radon tests should not be included as a condition of sale in real estate transactions."[25]

In contrast to the radon management people, he urged that government policy not be directed indiscriminately throughout the country, but "only where needed—namely in those areas where there is a strong potential of radon being present." His conclusion made clear his concerns about the adverse effects an aggressive policy might have on his group's financial interests. "To simply prescribe that all housing must incorporate radon prevention techniques would unnecessarily add to the cost of housing and further worsen the housing affordability problem."[26]

Rosengarten's 1990 testimony was less contained. His organization rhetorically supported the EPA's and Surgeon General's advice that was issued after his earlier testimony; that every home be tested. But he denounced "techniques that are alarming and designed to frighten the general public about the overall health risks posed by radon."[27] He criticized the EPA for "inexplicably [promoting] the idea that consumers could estimate their average annual exposure from a single test." He found "even more disturbing" the EPA's "sensationalistic ad campaign that frightens homeowners into taking actions they may not need to take and spending money they may not need to spend because of a false sense of alarm received from a single, false high result."[28] As the government's policy became more aggressive, the home building interests grew more agitated.

In 1991, on behalf of the NAHB, David Jackson reaffirmed the organization's hardened position. He criticized the EPA's "misplaced emphasis" on short-term tests as doing "a great disservice to the American public." He maintained that "there is no consensus among the scientific community regarding the health risks at various levels of radon exposure." While suggesting that long-term tests in selective areas might be appropriate, he raised doubts about the wisdom of any concerted radon testing policy without more knowledge. "It should be strongly emphasized," he said at the conclusion of his testimony, "that there is still much mystery surrounding the issue of radon."[29]

The other real estate organization to offer testimony at congressional hearings was the National Association of Realtors (NAR). In 1987, the organization issued a statement exhibiting a balance between concern for health and affordable housing. "We want to take an active role in attempts to reduce the potential health threat that radon gas may pose, while protecting the value and salability of homes."[30]

The next year, in 1988, Robert Ferguson testified on behalf of the NAR as Rosengarten had done for the National Association of Home Builders. The positions of the two organizations were virtually identical; they had joined to form a Real Estate Working Group with the EPA to discuss radon monitoring and mitigation techniques.[31]

Ferguson, like Rosengarten, suggested that policy be developed cautiously and that more research was necessary. He noted that radon had already had a "major impact" on the real estate industry insofar as transactions had been delayed because of testing, retesting, and mitigation questions. The NAR, he pointed out, also supported the allowance of radon mitigation costs as medical expense deductions for income tax purposes.[32]

Norman Flynn, the NAR's president, testified in 1990 that to force radon testing as a requirement of real estate transactions would reveal information about only 3 percent of all housing units per year. "So for us to, in fact, hit all the housing units, it would take us a generation to get that accomplished." He concluded that if radon is a "major national problem, it needs a national solution," not one that would single out real estate transactions as the locus of required testing.[33] Overall, his testimony maintained the tone offered by the NAR in earlier presentations to Congress: recognition of the importance both of health and affordability, and support for voluntary rather than required actions.

The most vigorous opposition to EPA radon policies at any of the hearings came from the National Apartment Association (NAA). Claiming to represent 200,000 real estate professionals involved in the multihousing industry, the association presented a written statement for inclusion in the 1990 hearing. While approving of nonregulatory efforts to inform the public about radon, it expressed concern that the EPA may have been distorting the health risks of radon and creating "unnecessary panic among the public."[34]

The NAA deplored the "unreasonable sense of urgency on the part of the EPA and the radon abatement industry. There appears to be a push to put standards in state and local building codes without adequate factual research about the nature of radon gas nor the true human occupant exposure risks." The association expressed concern that the EPA might be acting as a result of pressure from those who would make money from a more active policy.

> We encourage the EPA to continue its non-regulatory program, but it must also temper its public health messages with true risk-level statements. There is a feeling among building owners and managers that the promotion of radon as an imminent threat to health is a promotion by an industry that has much to gain financially.[35]

The apartment association questioned the validity of extrapolating from the uranium miner studies, one of the few times the question was raised at any of the hearings. It concluded by saying:

> NAA recognizes that radon is potentially a hazard to human health and believes EPA should promote efforts to mitigate and to continue their research efforts. However, it must be done in a responsible manner. To NAA, this means literature and media messages with clear responsible risk communication that is void of hysteria and extrapolation from studies that are not comparable.[36]

In sum, testimony by the financial losers was more diverse than by the gainers. The matter was simpler for the gainers. They could emphasize their interest in protecting people's health, with financial benefits being incidental to the noble purpose of their work. If financial losers spoke against a strong radon policy, however, they risked the opposite impression—that they were less interested in the public's health than in their own pocketbooks. While some financial losers were sensitive to this tension and tried to balance their presentations, others, notably the NAA, were directly critical of the EPA's aggressive approach.

Public Interest Groups

No matter how well intentioned, the motives of groups that have a financial stake in a policy are likely to remain suspect. Conversely, organizations with no apparent financial interest escape that shadow. Whether factually right or wrong, their ostensibly untainted views seem inherently more credible. But even supposedly disinterested groups may derive advantages associated with endorsement of a position or product.

In the fall of 1991, an EPA-sponsored publication listed 11 nonprofit organizations as "cooperative partners" in the agency's radon program. Some, such as the American Medical Association (AMA) and the American Lung Association (ALA), are health-oriented societies. Others are civic groups that had no explicit interest in health issues, such as the National Civic League, the National Association of Counties, and the Consumer Federation of America. Yet all were described as engaged in radon awareness programs, including sponsoring conferences and distributing materials on the subject.[37]

Unstated in the description was the fact that all the cooperative partners were receiving grants from the EPA to run their radon campaigns. Lisa Hucek, a spokeswoman for the agency, said that the EPA had encouraged these and other organizations to write grant proposals for their own radon projects. The EPA then made awards that ranged from $20,000 to $500,000.[38] Out-of-pocket expenditures by the organizations were nominal and usually amounted to little more than providing staff time.[39]

The EPA's awards did not necessarily dictate the organizations' positions. When prestigious associations endorse a policy position, however, they are likely to influence others, and financial arrangements relative to their advocacy should be made clear. A second question should be asked as well: What special base of knowledge does the organization possess that should allow others to be swayed by its position on the issue?

In summarizing the views of groups with no publicly visible financial interest in radon policy, three categories may be delineated: education and consumer groups, health and environmental groups, and nonindustry scientists. Each undoubtedly holds that its position was taken purely in the public interest. But other possible reasons should not be ignored. Beside money grants that some received, motivations could include desire to publicize and enhance the image of the association. Other incentives might relate more personally to spokespersons for the organizations: career benefits or ego gratification from publicity or congressional appearances.

Education and Consumer Groups

Several education and consumer groups enunciated positions on radon policies. Although none had obvious financial interests in the matter, some represented private or special interests. In 1987, for example, Kenneth F. Melley appeared before a congressional committee on behalf of the National Education Association (NEA) and the National Parents and Teachers Association (NPTA). He testified in support of proposed radon legislation to test and mitigate in schools. He worried that students and teachers were unknowingly being "exposed to a serious, silent, and deadly threat."[40]

His remarks were balanced by more temperate observations from William J. Leary, Jr., on behalf of the American Association of School Administrators (AASA). Leary also supported the bill under consideration, but was concerned that its wording "is somewhat biased toward the assumption that there is indeed a problem with radon in the nation's schools. AASA believes that the verdict is not yet in and that further study is needed." His organization felt that research thus far "does not support a basic assumption that there is a pervasive radon-in-schools problem."[41]

The school administrators association recognized that expenses for radon testing and mitigation might come from school budgets, and in that sense the interest of its constituents was related to finances. Leary noted that school budgets were already strained, and radon management could cost them more.[42] But his testimony, while moderately skeptical about the assumptions concerning the hazards of indoor radon, was sober and factual.

No representative from the AASA appeared at any subsequent hearing on radon. This was true even when the issue directly concerned schools, as in the 1992 hearing. In considering legislation that would require every school in the country to be tested, the only education representative to testify was Bob Chase on behalf of the National Education Association and the National Parents and Teachers Association.

Echoing the organizations' sentiments offered at the hearing 5 years earlier, Chase emphasized the need for more action on radon. Consistent with their dedication "to combat and eliminate environmental hazards in our nation's schools," the NEA and National PTA supported the proposed law. Chase recognized that the provision to test and mitigate schools could cost taxpayers more than $500 million. He offered "our staff and membership support in assisting" the effort; in other words, their goodwill.[43]

The principal consumers group to be heard at congressional hearings was the Consumer Federation of America, which claimed to represent some 240 local, state, and national organizations with more than 50 million people. On behalf of the federation, Susan A. Weiss testified in 1988 that the organization was committed to "decent and safe housing." Unlike other spokespersons, she focused her comments on the federal Department of Housing and Urban Development (HUD). She accused the department of failing to lead in "a national priority to reduce exposure to elevated levels of radon in the homes of all Americans." Virtually ignoring other agencies, she seemed to blame HUD exclusively—for burying "its head in the sand" and retreating "from responsibility for radon contamination."[44]

Mary Ellen Fise represented the Consumer Federation at two subsequent hearings and enlarged the focus of her organization's interest in radon. In 1991 she said that a survey conducted by the federation showed "widespread ignorance regarding the adverse effects of radon and the means by which it can be detected." In consequence, her organization believed "there is a need for an expanded federal response." This includes mandatory testing of all federally financed housing and "institutionalization of radon disclosure" for all residential real estate transactions.[45]

Fise testified again in 1992 in support of a bill introduced by Congressman Edward Markey. The bill would reauthorize assistance for state radon programs, require testing of all schools, and establish a proficiency program for testers and mitigators. The Consumer Federation of America supported these and other provisions of the bill, she said, because they address "two fundamental issues for consumers: saving lives and protecting consumers in radon-related transactions."[46]

Health and Environmental Groups

The American Lung Association offered a statement at the Senate hearing in 1991 that endorsed efforts to encourage universal testing. It urged enactment of legislation to broaden radon education pro-

grams, require tests for real estate transactions involving federal mortgage assistance, and mandate testing in schools throughout the country. The statement declared that education and consumer awareness
"cannot be emphasized enough." "The time is ripe," said the association, "for a national response to [this] public health threat."[47]

The Environmental Defense Fund (EDF) offered impassioned statements at hearings in 1985 and 1987. Claiming in 1985 that residential
radon was "one of the most serious environmental threats to public
health," it criticized the EPA's 4-picocurie action level as too lax. "It
would be a tragedy to lull the American people into a false sense of
security by leading them to believe that reducing radon to EPA's
target level will protect them."[48]

At the 1987 hearing, a representative of the EDF, Robert H.
Yuhnke, criticized the "wholly inadequate resources" committed to
radon reduction and called the EPA's exposure guideline "unconscionable." He called for enhanced monitoring, remediation, educational
programs, and a congressionally mandated goal to reduce indoor
radon "to natural background levels found in the ambient air."[49] But
the passion of his testimony was outdone by a spokesman for another
group dedicated to the radon issue, Jeff Koopersmith of the American
Radon Coalition for Health.

Koopersmith testified in 1990 but did not reveal the nature or
membership of his organization. A congressional aide told me it might
have been linked to radon technology companies, but his presentation
gave the appearance that his group was a public interest organization.
(Koopersmith was not asked about his organization at the hearing, and
he did not return several telephone calls.)

In his 1990 testimony, Koopersmith chastised the federal government for not forcing homeowners to test and mitigate. He argued that
failure to act was costing billions of dollars a year in health expenses.[50]
He submitted written testimony on behalf of his organization as well.
It was a single-spaced 63-page report, the longest included in any of
the hearings, and was filled with rhetoric. The following excerpt
provides the flavor.

> It is not only a tragedy to lose what EPA estimates as 21,000
> American lives each year, it is an impropriety to speak at one press
> conference after another and repeat these statistics time and again
> so as to transform them into irrelevancies. *Fifty to one hundred men,
> women and children* [original emphasis] may die of radon-related
> cancer every day. Yet we continue a radon program based on edu-
> cation and voluntary effort. No wonder Americans do not take our
> warnings seriously. The apparent hypocrisy of reading death tolls
> like these, compared to a set of programs that rely on some hope

that people will test for indoor radon just because they are interested in good health, is pointless and shrouds a crisis of national proportion in a curtain of burlesque.[51]

Koopersmith's written testimony concluded that "it is clear that the only way we can address this national crisis is either to provide, under a Federal program, testing and mitigation services to all Americans through grants, tax incentives, and interest-free loans, or mandate regulatory radon health standards."[52]

Koopersmith implicitly criticized the EPA for trying to make sure that radon management companies are proficient and honest. The agency, according to his report, "spends quite too much time, energy, and taxpayer dollars to protect citizens against a handful of unscrupulous companies at great cost to communications programs."[53]

The author in effect opposed vigorous efforts to detect unscrupulous companies, while favoring regulations that force homeowners to use these companies. Moreover, his implicit plea for more communications programs was at odds with his earlier statement that communications (education and information programs) have had little effect. The testimony was relentlessly ominous and hyperbolic. In the end it called for an impossible policy goal—to eliminate indoor radon everywhere.

> Now, armed with considerable data that radon is only increasing in complexity and health complications, we feel there is no choice but to weigh direct and authoritative means by which to eradicate radon from the American indoor environment.[54]

Scientists

The remaining nonprofit advocacy came from independent scientists. With the exception of John Harley of the National Council on Radiation Protection and Measurements (NCRP), a group chartered by Congress, the scientists spoke as individuals and claimed to represent no particular organizations. Nevertheless, each undoubtedly spoke for many in the scientific community.

At the first hearing in 1985, four scientists offered testimony. They were physicists with expertise in radiation, and all expressed the need for action. Harley referred to recent reports issued by his organization concerning risks from exposure to radon and its daughters. He said they indicated that two-thirds of the American population are exposed to radon daughter levels that increase the lifetime risk of lung cancer by 0.2 percent. He compared this with a lifetime mortality risk for lung cancer in smokers at about 10 percent.

Harley's testimony largely involved presentation of data with little policy advice. He did mention, however, that the NCRP recommended that a radon daughter concentration of 2 WLM (working-level month) per year be used as a guide for remedial action. This translated roughly into radon concentrations of 8 picocuries per liter of air, twice the action-level figure proposed by the EPA. He noted as well that achievement of a reported EPA goal to reduce lung cancer deaths from radon by 50 percent would be a formidable task.[55]

Bernard Cohen, a professor of physics and radiation health at the University of Pittsburgh, called radon "a national problem." He faulted the government's inadequate expenditure for radon research and control, noting that it was less than 1 percent of that for other sources of radiation like nuclear power.[56]

Richard Wilson, a professor of physics at Harvard University, expressed similar concerns. While recognizing that "there should be further research of the health hazard itself," he emphasized that even more necessary was research on exposures and risk management options. He suspected that policy would lead eventually to where "we are going to have to measure every house in the country."[57]

The last scientist to speak at the 1985 hearing was Anthony Nero of the Lawrence Berkeley Laboratory in California. He estimated that "something on the order of 1 million homes probably need remedial action, if you take as a remedial action level something in the vicinity of the level recommended by the NCRP." While saying there was no need for panic, he called for "a sense of urgency."[58]

Thus, the scientists largely agreed at this early date that radon constituted a hazard in many homes and that a national response was appropriate. At the same time, the EPA was credited with performing reasonably well in view of budgetary and political constraints. Doubts about the appropriate level of concern and the EPA's role were to come later.

Another congressional hearing at which independent scientists gave testimony took place 2 years later. In 1987, two scientists presented sharply contrasting positions. Gary Lyman, chief of the Moffit Cancer and Research Institute at the University of South Florida, was certain and uncompromising about what to do. He labeled radon "the single most lethal environmental carcinogen known to us today" and deemed the EPA's action level too permissive. "Efforts should be made to bring indoor radon levels in homes, schools, and other public buildings to levels as close to background as possible."[59]

Alan Hawthorne, head of the health studies section of the Health and Safety Research Division at Oak Ridge National Laboratory in Tennessee, was far less aggressive. Rather than naming radon a singularly alarming pollutant, he noted that "radon happens to be the indoor air pollutant currently drawing the greatest public attention. A few years ago, it was formaldehyde. Before that, it was asbestos." He emphasized the need for more research about "the issue of low-level exposures to complex mixtures." While affirming that radon reduction in homes with elevated concentrations need not await refinement of health risk estimates, he favored a restrained rather than aggressive policy.[60]

Of the independent scientists who appeared before congressional committees, all presumed that indoor radon at some level should be addressed. Except for Lyman, none counseled extreme positions. Indeed, by 1991 some of the scientists had retreated from the positions they espoused in earlier testimony. Bernard Cohen now believed that the billions of dollars necessary to mitigate lower concentrations in homes was a waste. His investigations had convinced him that "low-level exposure is essentially harmless; there is no effect at very low levels, as are found in most homes."[61] Similarly, Anthony Nero's work persuaded him that perhaps 100,000 homes in the United States contained radon levels that offered significant lifetime risks to residents. This was not only radically lower than the EPA's 8 million figure, but also much lower than his own 1985 estimate of 1 million.[62]

After 1985, Cohen was not invited to testify at any hearings, and Nero appeared at only one other. Indeed, most subsequent hearings on radon were conducted without any witnesses from the scientific community.

Sorting Reasons

In reviewing the contribution of varying interests to the development of radon policy by way of congressional testimony, several points are clear. Virtually everyone who testified considered indoor radon a potential health risk, though of varying degrees of seriousness. In this era of heightened concern about the environment and health, to suggest less caution than more seems a callous gamble with people's lives. Advocating an aggressive policy is therefore likely to be seen as in the public interest and to be more popular. Concerns about popularity can never be far from an elected official's thoughts.

Except for some in the real estate or building industries, interest groups tended to fall in line behind a vigorous radon policy. While the positions of financial losers and gainers balanced each other, the nonprofit advocacy associations tilted in the direction of a strong policy. No countervailing "disinterested" groups offered testimony.

Why the dearth of nonprofit groups who would challenge the aggressive position on radon? Apart from the general wish not to create risks for people, there is the nature of the issue. A layperson is less likely to be confident about an issue associated with scientific complexity. A greater challenge from nonscientists would be easier if more scientists spoke out publicly. Although many scientists were skeptical about the EPA's approach, few assumed a publicly adversarial position.

A compounding element is the fact that few scientists were invited to testify after the 1985 hearing. Spokespersons for the Consumer Federation of America, the National Education Association, and other groups without particular knowledge about the subject were invited far more frequently. Indeed, at five of the subsequent eight hearings, not one scientist was present.

This is not because skeptical scientists are unavailable. There are scientists aplenty inside government and out who are lopers and loppers (supporters of a more limited radon policy or none at all). They are in the Department of Energy, the National Cancer Institute, the National Council on Radiation Protection and Measurements, the Health Physics Society, and research and academic institutions across the country. Yet rarely were any invited to speak at congressional hearings. Thus, views by these members of the scientific community were hardly ever heard by the lawmakers.

Contrary to this pattern, two scientists were invited to testify at the congressional hearing in 1992. The proceedings were instructive. The hearing was on the radon bill introduced by Congressman Markey. The bill provided for additional radon grants to the states, mandated a national certification program for radon testing, and required that every school in the nation be tested. The hearing was the ninth on radon by a congressional committee since 1985. As at the other hearings, most witnesses supported strong radon policies.[63]

After Markey and EPA spokesman Michael Shapiro testified, the remaining eight witnesses took seats next to each other at the witness table. Following their prepared statements, Representative Al Swift, the chairman, asked how each would vote on the bill. Six said they would support it without qualification. Two said they recognized that radon at some level could cause lung cancer, but they declined to endorse the bill. These two were the only scientists at the table.

Jonathan Samet, an epidemiologist from the University of New Mexico, felt that a public policy response to radon was warranted, but he was not sure the bill provided the best approach. The other scientist, Jan Stolwijk, an epidemiologist at Yale University, also demurred from endorsing the bill.[64] He had just offered testimony emphasizing that a reduction of cigarette smoking would save more lives than reduction of indoor radon.[65]

Chairman Swift lightheartedly observed that the scientists were "acting like scientists" and made clear his support for the wisdom expressed by the other six witnesses. The six included two from radon testing and measurement companies; spokespersons for the National Education Association, the Consumer Federation of America, and the Environmental Law Institute; and an official from the New Jersey Department of Environmental Protection and Energy.

Unstated at the hearing was the fact that the six witnesses represented organizations or agencies that were receiving money or program assistance in connection with their positions. The private companies profited directly and, with the exception of the National Education Association, the others had received grants from the EPA to promote radon awareness programs. (The EPA also paid for and published more than 100,000 radon information brochures that carried the imprimatur of the National Education Association and the National Parent Teacher Association.) Sincere as the witnesses may have been, the fact that their organizations were EPA grant recipients deserved recognition.

The character of the hearing was typical of the others held on radon over the years. The exchanges between lawmakers and most witnesses amounted to perpetuation of a shared belief. By inviting witnesses who tended to be leapers or loopers (supporters of aggressive policies), members of Congress received limited exposure to lopers and loppers.

This is not to say that those who gave testimony were venal or wrong. "Right" or "wrong" are judgment words that will make sense only in hindsight, after more scientific evidence becomes available. But over the years, experts who could have presented countervailing views were infrequently called to testify. This almost surely influenced the perceptions of the lawmakers and the consequent legislative agenda.

If Congress fails to summon representatives who can offer the full range of views on a subject, another American institution presumably is in place to do so—the media. The next chapter assesses the role of the press in the radon debate.

Notes

1   Karl W. Deutsch, *Politics and Government: How People Decide Their Fate*, 3d ed. (Boston, MA: Houghton Mifflin Co., 1980), 10.
2   Anne L. Hiskes and Richard P. Hiskes, *Science, Technology, and Policy Decisions* (Boulder, CO: Westview Press, 1986), 183.

3   Gabriel A. Almond and Sidney Verba, *The Civic Culture* (Princeton, NJ: Princeton University Press, 1963), 300.

4   Leonard A. Cole, "Cashing in on Confusion," *New Jersey Monthly*, Vol. 16, No. 8 (March 1987), 23.

5   House Subcomm. on Health and the Environment of the Comm. on Energy and Commerce, *Hearing on Radon Exposure: Human Health Threat*, Nov. 5, 1987 (Washington, DC: Government Printing Office, 1988), 2, 124.

6   Ibid., 126–27.

7   Senate Subcomm. on Superfund, Water, and Ocean Protection of the Comm. on Environment and Public Works, *Hearing on Radon Testing for Safe Schools Act*, May 23, 1990 (Washington, DC: Government Printing Office, 1990), 111.

8   House Subcomm. on Health and the Environment of the Comm. on Energy and Commerce, *Hearing on Indoor Air Pollution*, Apr. 10, 1991 (Washington, DC: Government Printing Office, 1991), 229.

9   House Subcomm. on Natural Resources, Agricultural Research and Environment of the Comm. on Science, Space, and Technology, *Hearing on Federal Efforts to Promote Radon Testing*, May 16, 1990 (Washington, DC: Government Printing Office, 1990), 84.

10 Ibid., 85–86.

11 Senate Subcomm. on Superfund, Ocean, and Water Protection of the Comm. on Environment and Public Works, *Hearing on Pending Radon and Indoor Air Legislation*, May 8, 1991 (Washington, DC: Government Printing Office, 1991), 21–23, 60–63.

12 Ibid., 62–63.

13 Ibid., 63.

14 Ibid., 69.

15 House Subcomm. on Transportation and Hazardous Materials of the Comm. on Energy and Commerce, *Hearing on Radon Awareness and Disclosure*, June 3, 1992 (Washington, DC: Government Printing Office, 1992), 44.

16 Ibid., 43

17 National Research Council, *Comparative Dosimetry in Mines and Homes* (Washington, DC: National Academy Press, 1991), 4.

18 *Hearing on Radon Awareness and Disclosure* , June 3, 1992, 44–45.

19 Gun Astri Swedjemark and Astrid Mäkitalo, "Recent Swedish Experiences in Radon-222 Control," *Health Physics*, Vol. 58, No. 4 (Apr. 1990), 454.

20 House Subcomm. on Natural Resources, Agriculture Research and Environment of the Comm. on Science and Technology, *Hearing on Radon and Indoor Air Pollution*, Oct. 10, 1985 (Washington, DC: Government Printing Office, 1985), 98. The Reading Prong extends through Pennsylvania, New Jersey, and New York.

21 Ibid., 98–99.

22 *Hearing on Radon Exposure: Human Health Threat*, Nov. 5, 1987, 109.

23 Ibid., 112–14.

24 Senate Subcomm. on Superfund and Environmental Oversight of the Comm. on Environment and Public Works, *Hearing on Radon Contamination: How Federal Agencies Deal with It*, May 18, 1988 (Washington, DC: Government Printing Office, 1988), 134.

25 Ibid., 135–36.

26 Ibid., 139–40.

27 *Hearing on Federal Efforts to Promote Radon Testing*, May 16, 1990, 105.

28 Ibid., 105–6.

29 *Hearing on Pending Radon and Indoor Air Legislation*, May 8, 1991, 20–21.

30 *Hearing on Radon Exposure: Human Health Threat*, Nov. 5, 1987, 139.

31 *Hearing on Radon Contamination: How Federal Agencies Deal with It*, May 18, 1988, 35–37.

32 Ibid.

33 *Hearing on Federal Efforts to Promote Radon Testing*, May 16, 1990, 184.

34 Ibid., 226.

35 Ibid., 227–28.

36 Ibid., 229.

37 *Radon Bulletin*, Conference of Radiation Control Program Directors and the U.S. Environmental Protection Agency, Vol. 2, No. 1 (Fall 1991), 6–9.

38 Interview, June 2, 1992.

39 Description of involvement by organizations came from Jeffrey Boal of the Advertising Council, interview, Feb. 12, 1992; William Hendee, former vice-president of the American Medical Association, interview, May 4, 1992; and Paul Locke of the Environmental Law Institute, interview, June 2, 1992.

40 Senate Subcomms. on Environmental Protection, and Superfund and Environmental Oversight, of the Comm. on Environment and Public Works, *Hearing on Radon Gas Issues*, Apr. 2, 1987, (Washington, DC: Government Printing Office, 1987), 88.

41 Ibid., 95.

42 Ibid., 98.

43 *Hearing on Radon Awareness and Disclosure*, June 3, 1992, 55.

44 *Hearing on Radon Contamination: How Federal Agencies Deal with It*, May 18, 1988, 37–38.

45 House Subcomm. on Health and the Environment of the Comm. on Energy and Commerce, *Hearing on Indoor Air Pollution*, Apr. 10, 1991 (Washington, DC: Government Printing Office, 1991), 197, 203–4.

46 *Hearing on Radon Awareness and Disclosure*, June 3, 1992, 104.

47 *Hearing on Pending Radon and Indoor Air Legislation*, May 8, 1991, 95–96.

48 *Hearing on Radon and Indoor Air Pollution*, Oct. 10, 1985, 282.

49 *Hearing on Radon Exposure: Human Health Threat*, Nov. 5, 1987, 65, 75–78.

50 *Hearing on Federal Efforts to Promote Radon Testing*, May 16, 1990, 116–18.

51 Ibid., 119.

52 Ibid., 121.

53 Ibid., 142.

54 Ibid., 181.

55 *Hearing on Radon and Indoor Air Pollution*, Oct. 10, 1985, 186–87.

56 Ibid., 161–62.

57 Ibid., 180–85.

58 Ibid., 204–5.

59 *Hearing on Radon Exposure: Human Health Threat*, Nov. 5, 1987, 88–89.

60 Ibid., 97–98.

61 Jane E. Brody, "Some Scientists Say Concern Over Radon Is Overblown," *New York Times*, 8 Jan. 1991, C-4.

62 Ibid.
63 *Hearing on Radon Awareness and Disclosure Act of 1991*, June 3, 1992.
64 Ibid., 116–17.
65 Ibid., 81–82.

CHAPTER 8

The Press, Science, and Radon

"SCIENTISTS ARE TO REPORTERS WHAT RATS ARE TO SCIentists." With these words, Victor Cohn, science reporter for *The Washington Post*, implied that the press scrutinizes science and scientists with the thoroughness that scientists are presumed to investigate their research subjects.[1] The analogy, though amusing, is overdrawn. The press's coverage of science has often been criticized as less critical than its coverage of other activities.

Reporting Science

The reporting of politics, economics, art, and literature is commonly analytical and critical. Not so with science, which is more often presented as explanation. Compared with public servants or entertainment figures, scientists receive a "light dose of press scrutiny," Rae Goodall found.[2] According to Dorothy Nelkin, journalists explain science and technology subjects and rarely challenge their sources of information.[3] Stephen Klaidman wrote that while reporters may understand the nonscientific aspect of health-risk stories, "they are often at a loss when it comes to reporting on the science."[4]

The consequence of inadequate media coverage involves more than a question of rectitude. The media not only help shape public concerns; they influence the policy agenda of decision makers. If the press fails to analyze critically, the basis for policy may suffer. When a New Jersey state senator was asked if press coverage stirred his interest in radon, he replied, "Yes, I mean that's what stirs my awareness of virtually everything."[5]

The media are not always reluctant to question the conduct of scientific and technological activities. The wisdom of spending billions of dollars for projects like the superconducting supercollider and human genome mapping program have been critically reported in the national media. This has been especially true as the huge costs of "big science" projects cut the availability of funds for smaller projects.[6]

In addition, calamities born of scientific and technological miscalculations have received broad coverage. The Chernobyl and Challenger catastrophes and the troubles associated with the Hubble space telescope have been extensively reported. But criticism of these events will not likely influence the press's traditionally passive approach to less spectacular scientific activities. The underlying reasons lie in the way science is seen by the public.

Distinctiveness of Science

Why the more deferential approach to science than other areas? One reason is science's perceived complexity. Reporters, like most people, consider politics and art within their grasp. The operations that underlie these disciplines—their rationales, techniques, and consequences—appear understandable even to observers who lack the talent to perform them. But the building blocks of modern science are another matter. They frequently involve mathematically based theories and sophisticated laboratory devices. Nonexperts consider them too complex to grasp. Most reporters are evidently reluctant to probe areas whose bases appear beyond their comprehension.

The second differentiating characteristic about science is power. Power is also central to politics, though in a simply understood form: influencing people to behave in certain ways. The power associated with science reaches beyond human interaction. Science offers the potential to alter the human condition in ways that no other discipline can. The power of the atom or of genetic engineering allows for staggering destruction or irreversible change of human life. It is more profound and global than power associated with other human activities.

Third, science bears a magical quality. It commonly contradicts expectations born of habit. When two colorless liquids are mixed and instantly become lustrous red, the phenomenon defies everyday experience. The uninitiated observer witnesses a seemingly inexplicable event. Happenings at the edge of the universe or in atomic nuclei appear more mysterious than intelligible. A humanities graduate student recalled to me her shock when she was a sixth grader and her teacher said that all matter is composed of tiny nuclei surrounded by whirling electrons. She was so disturbed by the contradiction to her senses that she has felt anxious about science ever since.

Finally, science is ostensibly based on essential truths that are not inherent in other enterprises. While these "truths" are subject to modification as new discoveries are made, at any one time they are accepted as the root explanations for the workings of nature. They carry a cache of verity that is absent from other disciplines. Religion, while making claim to eternal truth, is grounded in faith and unprovable belief. Science seeks predictable, replicable explanations. No adjective offers the weight of verity more than the word "scientific." In other disciplines—economics, politics, art, literature—a point is enhanced when describing it as scientifically valid. By contrast, no scientific fact is strengthened by calling it economically true, politically true, artistically true, or literarily true.

Radon and the Press

In considering the indoor radon issue, a journalist may hesitate to challenge the prevailing orthodoxy for reasons beyond the usual circumspection about scientific subjects. No responsible person wants to encourage people to ignore authoritative advice to correct a hazard. Moreover, the government has been frequently criticized in the past for understating dangers prompted by scientific and technological activities. The government imposed controls over a variety of substances long after their damaging consequences had become evident. This was true for pesticides and other chemicals, tobacco smoke, and radiation associated with nuclear-processing activities.

Environmental devastation to lakes and forests caused by acid rain, for example, has been documented for years. Researchers found the major source of acid rain to be from industrial and other human activities. Chemicals released in smoke from industrial plants rise and acidify moisture in the upper atmosphere. In the form of rain the moisture falls to Earth and poisons the lakes and trees. Not until 1989, after cumulative research by hundreds of scientists at a cost of one-half billion dollars, did the government acknowledge the severity of the problem and seek a substantive policy.[7]

In the case of radon in homes, the approach has been reversed. Every homeowner has been urged to test and remediate *before* epidemiological evidence could be found that residential radon is a widespread hazard. This nonfinding was scarcely discernible from press reports, however. Despite uncertainties that were built into news stories about radon, the media usually echoed the official government line as if the evidence were beyond dispute.

Journalists showed their traditional reluctance to question scientific experts. Countervailing interest groups who challenged the official wisdom were largely from the real estate industry. Since they had an economic stake in the matter, their skepticism was treated with suspicion. Few reporters seemed prepared to challenge the official orthodoxy. But newspapers and the other media should have been asking the questions that have been raised in this book.

Trends in Press Coverage

How did the press cover the radon issue? This examination focuses on the frequency and content of newspaper articles. *The New York Times* had many more articles on the subject than other national newspapers, including *The Wall Street Journal*, *The Washington Post*, *The Los Angeles Times*, and *The Christian Science Monitor*.[8] In part this was because the *Times* provides extensive coverage of New Jersey, where the radon issue was highly salient. Two New Jersey newspapers, *The Star-Ledger*, published in Newark, and *The Record*, published in Hackensack, were also screened for articles on radon. *The Star-Ledger* is the state's largest circulation newspaper and the principal daily covering the middle of the state. *The Record* is the major newspaper in northern New Jersey. Both serve locations in the state in which radon became a significant issue of local concern.

The quantitative trend of radon-related articles in *The New York Times*, *The Star-Ledger*, and *The Record* reveals that attention grew in the mid-1980s, but fell by the end of the decade.[9] As the figure below shows,

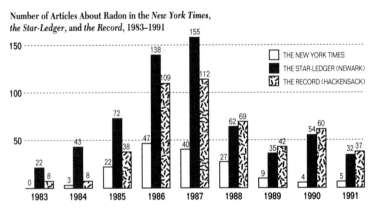

Number of Articles About Radon in the *New York Times*, the *Star-Ledger*, and *the Record*, 1983–1991

Figures for *The New York Times* are calculated from the annual *New York Times Index* under the listing for "radon gas." Figures for *The Star-Ledger* and *The Record* are calculated from annual data bases for those newspapers under the heading "radon."

The Star-Ledger published the most radon items each year through 1987, followed by *The Record*. This probably related to the content orientation of the two papers, which will be discussed below. But the most striking indication from the graph is the peak reached in 1986 and 1987 by all three papers, followed by sharp declines in the subsequent years. The number of radon articles in the *Times* rose from 0 in 1983 to a high of 47 in 1986, and a decline to 5 in 1991. In *The Star-Ledger*, the numbers were 22 in 1983, a peak of 155 in 1987, and a fall to 32 in 1991. In *The Record* there were 8 in 1983, a high of 112 in 1987, and a decline to 37 in 1991. The numbers in the mid-1980s coincided with newly announced government concerns, particularly in the New York–New Jersey–Pennsylvania region.

While newspaper coverage declined by the end of the decade, the trend was paradoxical. In 1988 and 1989, the government was issuing increasingly pronounced warnings that as many as 20,000 people were dying annually of radon-caused lung cancer; virtually all homes needed to be tested for radon; and remediation should be undertaken if concentrations exceeded 4 picocuries. Indeed, by 1991 the EPA had raised its uppermost estimated figure to 30,000 annual lung cancer deaths from radon. But as the government's warnings became more alarmist, the number of articles continued to decline.

Why the dip? One apparent reason is that the press no longer saw the subject as "news." The novelty of the finding had worn thin, and reporters were reluctant to repeat themselves. Yet this explanation ignores the new perils pronounced in later government messages. The most dramatic proclamations about radon came in 1988 and 1989. In the fall of 1988, Vernon J. Houk, assistant surgeon general with the Public Health Service, announced that new evidence showed that the threat of indoor radon was more widespread than previously thought. In urging the testing of almost all homes in the United States, he declared that "radon-induced lung cancer is one of today's most serious public health issues."[10]

Official warnings to the public during the next year were even more charged. Christopher Daggett, Commissioner of New Jersey's Department of Environmental Protection (DEP), called radon "the most serious environmental health threat to New Jersey residents."[11] William K. Reilly, administrator of the federal Environmental Protection Agency (EPA), proclaimed that "radon is one of the most serious environmental health problems we have today."[12]

Despite these statements, the number of newspaper articles on the subject continued to fall after 1987. Apart from the question of novelty, the explanation related to a sense of bad-news saturation. After the initial shock was aired that radon was a widespread hazard, the gas took its place in a seemingly endless parade of newly discovered dangers. During the past two decades, one commonly used item after another—pesticides, fertilizers, foods, drug products—were suddenly declared a menace to human health.

If reporters and the public suffer from hazard overload, the matter has become even more complicated. Some initially presumed hazards later seemed less of a problem. The extent of harm caused by cholesterol in the diet, mercury in fish, and asbestos in nonpeeling walls has become more equivocal in later consideration than when first recognized as potential problems for humans. In this perspective, a public that demurs from jumping in response to the government's advice to test for radon (and, if above 4 picocuries, to remediate) is more understandable.

Public inaction may also be attributed to the inconsistencies in official pronouncements. While reporters may not have adequately probed the premises of radon policies, the weaknesses could have been apparent to discerning readers. This was palpably true in New Jersey, where environmental and health authorities were unsuccessfully trying to address the Montclair radon-soil problem (see Chapter 6). At the same time these officials were urging citizens throughout the state to follow their advice to test and remediate.

Editorial reaction by *The New York Times* to the radon issue was more limited than that in the other newspapers. The only *Times* editorial on the subject appeared in September 1988, calling radon in the home "a serious and widespread hazard." At the same time it acknowledged that critics who think the 4-picocurie standard is unnecessarily stringent "may well be correct." The editorial ultimately came down in support of the EPA's action level because "it seems to make good sense in this case to err on the side of caution."[13]

A few weeks after *The Times'* editorial advice, the paper published an op-ed piece I wrote that questioned the EPA's policy. The prominent place given to the article, and conversations with staff members of *The Times*, convinced me that some of them viewed the EPA's policy more skeptically than was represented in the previous month's editorial. In the op-ed article, I questioned the wisdom of the EPA's aggressive policy, as I have in this book.[14] Then, as now, no study had found a statistically significant relationship between illness and radon in homes. No other editorial or op-ed comment in *The Times* has dealt with radon since then.

Radon stories appeared far more frequently on the news and editorial pages of the two New Jersey newspapers. By focusing in particular on their coverage of the Montclair soil issue, it will be seen that the different emphases each gave to the Montclair matter was emblematic of their approaches to the overall radon issue.

The *Star-Ledger* News Stories

Between 1983 and 1991, *The Star-Ledger* carried about 600 items on radon compared to nearly 500 in *The Record* and about 150 in *The Times*. As noted, the frequency pattern was similar for the three papers, peaking in 1986 and 1987, and sharply falling afterward. But content imbalance was more striking in *The Star-Ledger* than in the other two.

The Star-Ledger serves the Montclair area that confronted the remediation problem. Moreover, many of its readers live in areas designated by the EPA and DEP as high in natural radon concentrations because of geological formations.

New Jersey's first foray into remediation efforts began after the discovery that a small cluster of homes in Essex County had elevated levels of radon. The homes straddled the borders of the adjacent towns of Montclair, Glen Ridge, and West Orange and were initially thought to number 27. The story appeared during the first week of December 1983. *The Star-Ledger* reported that Governor Thomas Kean had signed an executive order giving emergency powers to environmental and health authorities to deal with the problem. According to the newspaper, "[s]tate officials said they expect to have the remedial plans designed within a week and the cleanup completed by year-end"—that is, by the end of 1983, in 4 weeks![15]

By early 1984, no action had been taken, and new proclamations from state officials were duly reported by the newspaper. Steven G. Kuhrtz, director of the DEP's Division of Environmental Quality, declared in February that a cleanup program would start within weeks and continue through the summer.[16] When at the end of the year no cleanup had begun, he could only tell a reporter that he did not expect a decision on the matter "this week."[17]

The soil was dug from around some homes in mid-1985 and placed in steel drums, which were left next to the homes. During the next 3 years, *The Star-Ledger* reported the series of unfulfilled disposal plans to move the drums. The paper also noted the plight of the four families who had moved out in 1985 and were still living in "temporary" quarters in 1990.

At the same time, hundreds of other area residents remained concerned that they were living in a situation deemed hazardous by state officials who seemed not to know how to alleviate the problem. In August 1989, in a plea to the EPA for help, Samuel Pinkard, head of the Montclair Radon Task Force, was quoted in *The Star-Ledger* as calling the affected area "the only site in New Jersey which is a superfund site where people are living on it."[18]

The Montclair problem was only one area of confusion about radon policies that could be gleaned occasionally from the pages of *The Star-Ledger*. In January 1989, for example, a headline indicated "'Surprising' Change in DEP Policy on Radon Testing Sows Confusion." The story referred to a statement by DEP Commissioner Christopher Daggett that the state would no longer confirm initial readings if above 4 picocuries, but only if greater than 8 picocuries. Several municipal health officers, including DEP Assistant Commissioner Donald Deieso, said they were unaware of the changed policy, according to the report. DEP officials estimated that the change would save $50,000 from the annually budgeted $1.2 million for such testing. Moreover, it would allow DEP staff to focus on other radon activities.[19]

Jorge Berkowitz, the department's director of environmental quality, said that while "anything under 8 picocuries becomes less important," homeowners should still see 4 picocuries as "the number to shoot for." Municipal officials responded in frustration that people did not know now at what point they should be concerned.[20]

But articles in *The Star-Ledger* suggesting uncertainty or confusion were comparatively rare. Typical news items contained unqualified statements of certitude, like Daggett's proclamation that radon was New Jersey's number one environmental health problem. "When I say that radon is the most serious health threat, I'm talking to all residents of New Jersey, not just the northern part of the state."[21]

The newspaper never seriously inquired into the validity of such statements. Rather, news stories through the 1980s echoed the official view that natural radon formations in New Jersey were a "potential hazards crisis"[22] and that "study after study has shown that the nation's radon problems are...widespread."[23]

The Star-Ledger Editorials

The Star-Ledger's editorials from the onset of the radon issue in the early 1980s reflected the establishment position. Moreover, they consistently defended the New Jersey DEP's actions in Montclair. In December 1983, the newspaper editorially praised the state's environmental authorities for finding the "dangerous relic of pollution" produced by radon in the Montclair area and lauded Governor Thomas Kean for creating an emergency cleanup program for the affected homes.[24]

Two years later, after some of the soil had been packed in drums, no one knew what to do next. Several DEP plans for permanent disposal had failed to materialize. But the newspaper absolved the department from fault because when it ordered the digging, it "believed it had resolved the disposal problem."[25]

The following year, in October 1986, while the drums still sat on the lawns with no place to go, *The Star-Ledger* could find no harsher word to describe the situation than "regrettable." The closest it ever came to criticizing the department's radon policy was to call the management of the Montclair soil "not...among the DEP's finest achievements."[26]

In another editorial on the subject, in May 1987, the newspaper attributed the soil problem to "bureaucratic snafus" and "communications blunders." Resolutely failing to criticize the DEP directly, the editorial spoke in the abstract of a "problem" that "continues to defy a solution."[27] Four months later, the newspaper repeated its no-fault theme. "The radon issue has proved to be more difficult than at first anticipated."[28]

At the end of 1988, the DEP was finally able to arrange for disposal of the excavated soil, and the drums were carted away. But the owners of the affected homes were still prohibited from returning because questions were unresolved about the safety of the remaining deeper soil. *The Star-Ledger* seemed unfazed by the lack of resolution and, without qualification, heaped praise on the DEP's radon program. Ignoring the Montclair issue and uncertainties about overall radon policies, it praised the precocious behavior of the state's environmental and health authorities.

> The wisdom of New Jersey officials in taking the [radon] problem seriously and devising a program to meet the challenge—at a time when the federal government had not yet become aware of the magnitude of the risks—is now demonstrated. It behooves those New Jerseyans who have not yet tested their residences for the presence of radon to wake up to the severity of the menace and take whatever steps are necessary to combat this environmental peril.[29]

From 1983 through the end of the decade, the paper's news pages chronicled the state's mishandling of the radon problem in several communities. Yet whatever the blunder of the moment by state officials, *The Star-Ledger's* editorials never questioned the state's radon policies. In few news stories and in no editorials did the paper allow that the premise of the DEP's or EPA's policies might be unwarranted. While reporting the frustrations in the Montclair area, and despite brief mention that epidemiological studies were inconclusive, *The Star-Ledger* presented the DEP's excuses and pronouncements without challenge.

The tilt is graphically illustrated by the absence of a response to a 1988 epidemiological investigation. In July, the state Department of Health (DOH) released the findings of a study of the residents of homes in Montclair, Glen Ridge, and West Orange where radon levels were elevated. The newspaper reported the findings in an inside page under the headline: "No Sign of Cancer 'Epidemic' in 3 Radon-Tainted Towns." The article noted a DOH official's acknowledgment that both the overall death rate and cancer rate were "below expectations."[30]

Radon policy is based, after all, on the presumption that residents in homes with higher radon concentrations suffer a higher incidence of lung cancer. Yet the contrary implication of the study's findings went unexplored in *The Star Ledger's* news and editorial pages.

The Record News Stories

While *The Star-Ledger* largely failed to criticize state actions concerning radon, *The Record* portrayed both the Montclair soil problem and overall radon policy with more ambivalence. This was especially true in the earlier years of the matter.

On December 2, 1983, *The Record* reported that residents and local officials of Montclair and Glen Ridge were "furious." They had just learned that "dangerous levels of radioactive gas" had been seeping into homes near the towns' common border. They heard about the problem not from the DEP, but from media reports. The department had known about the elevated radon levels from tests it conducted months earlier, but kept the information to itself. Local anger was fueled by the refusal of DEP Commissioner Robert E. Hughey and other state officials to disclose the level of radon in any of the homes.[31]

Apart from the manner that the information became public, *The Record's* news story included comments by state officials that seemed contradictory. Governor Kean announced that the state was "making every effort to deal with the problem as fast as possible." At the same time, the state's deputy commissioner of health, Dr. Allen N. Koplin, assured the public that there was no hurry; that "precipitous action is not warranted." The initial silence by state officials, now replaced by contrasting emphases, prompted town council members to castigate the state's behavior as "irresponsible."[32]

Two days after the news broke, *The Record* reported that state officials had met with residents of the affected area and assured them that the radon in their homes "did not present a serious health hazard." Nevertheless, the officials announced, testing and ventilating systems would immediately be installed in homes with the highest readings. The officials stressed that no evidence of excessive cancer existed in the area, yet they intended to remove the contaminated soil. They speculated that the contamination was caused by dirt that had been moved decades earlier from the defunct U.S. Radium Corporation. The plant had operated in the neighboring town of Orange, but closed in 1929.[33]

By the end of the week, the newspaper's reporters were citing experts who were divided over the danger of radon. Some were contending that it presented a major threat, while others thought it insignificant. The DEP also said that it found 27 area homes with elevated readings, 5 of which had levels above 100 picocuries per liter of air. Some officials expressed alarm at the findings, but DEP Commissioner Hughey insisted that no one could define the health risks to his satisfaction.[34]

On December 22, 3 weeks after the Montclair radon scare began, *The Record* reported that Dr. William Parkin, an epidemiologist with the New Jersey Department of Health, found no evidence of a cancer problem in the area. Among the 142 people who had been living in the area, several for more than 50 years, 5 had suffered incidences of some kind of cancer (including cancer of the breast, lung, colon, ovaries, or skin). The five cancers were fewer than would be expected if radon exposure had an effect, according to Dr. Parkin. The paper did not think the revelation worthy of front-page attention, however, and unlike many of the radon stories of the past weeks, it appeared in an inside section.[35]

In the same article, the DEP was cited as suggesting that more homes might have higher radon concentrations. It announced that in at least one house, radon levels were brought below outdoor levels after installation of a new furnace and venting system, but that correcting other homes might require soil removal. All houses in the area would be corrected within 18 months, said Richard Dewling, a regional EPA administrator and future commissioner of New Jersey's DEP.[36]

In June 1984, *The Record* reported a DEP announcement that in the summer it would begin excavating the soil. The cost would be between $15 and $60 million, according to James Staples, a DEP spokesman, who said he was "not sure if people will have to be [temporarily] relocated. Possibly."[37]

Plans for mass excavation moved ahead. The observation by the DOH epidemiologist 6 months earlier, which noted that the rate of cancer in the area was unusually low, seemed forgotten. Neither in *The Record's* news stories nor in editorials were questions raised about the basic scientific-health premise before excavation was undertaken. The DEP's estimate that the cost could be as high as $60 million seemed to faze no one.

Instead, the news reports began to focus on how to dispose of the contaminated soil. On this issue, as with earlier ones, the DEP seemed unsure. In the fall of 1984, the department announced that the soil would be moved to a national guard armory in nearby West Orange. As the town was preparing to challenge the DEP's plan in court, Commissioner Hughey announced a new plan. The department, he said, would begin excavating in January 1985 and truck the soil 3,000 miles to Washington State. Immediately afterward, DEP spokesman James Staples said he was unaware that the plan for local storage had been altered. *The Record* reported his concern that sending the soil to Washington would cost "oodles more money."[38] Neither he nor the commissioner would estimate how much the "oodles" actually would be.

Thus, the newspaper amply reported confusion and contradiction in the state's approach to the Montclair problem. Its news pages also carried rare reports that questioned overall radon policy as well. For example, Naomi Harley, a radiation physicist at the New York University Medical Center, had been trying to measure indoor radon levels since 1981. She told *The Record* in 1985 that finding tremendous variations in concentrations at the same location over time were "really driving me crazy." She talked of "eight or nine factors involved in radon entry" that constantly change—including soil moisture, barometric pressure, wind, and house temperature. While the conventional belief is that indoor levels are highest in winter when windows and doors are closed, she found higher readings in the summer and spring. "There are many surprises," she said. "Many of the things you think should happen won't happen."[39]

In 1988, *The Record* carried an article that cited a local mayor's skepticism about the danger of radon. He was reminded of "certain foods that people say aren't good for you, and then 6 months later they say it's O.K. I haven't heard anything concrete about radon being a problem here."[40] Soon after, another article indicated that several state agencies and the U.S. Department of Energy (DOE) questioned whether the recent EPA's advisories had been exaggerated. Susan Rose, the DOE's radon research manager, was quoted as saying that "radon risk is uncertain enough to warrant a major federal research commitment before widespread testing is urged for virtually every homeowner in the country."[41]

But items like these that raised questions appeared infrequently compared to those that echoed the establishment position. News stories abounded with unchallenged pronouncements that many homes contained "potentially dangerous amounts of radon";[42] that the gas "is a major national health threat";[43] and that radon is "seriously threatening the health of homeowners across the country."[44]

The Record Editorials

The Record's editorials initially reflected concern, while acknowledging ignorance. In October 1985, in its first editorial on the radon issue, the newspaper conceded that much about radon—its whereabouts and health effects—was unknown. At the same time the paper called radon in homes a "major health worry," "frightening," and "an environmental powder keg." The message seemed inconsistent. After admitting ignorance about the danger of residential radon, to describe it as a powder keg appeared unwarranted.[45]

The uncertainties about radon reported in *The Record's* news pages were reflected in inconsistent editorial messages. This was especially true in the case of the Montclair soil issue. Unlike *The Star-Ledger*, *The Record* did not view the DEP as blameless in the matter. Indeed, the paper seemed schizophrenic as its editorials alternately praised and criticized state policies.

In 1987, for example, *The Record* published eight editorials on radon, most dealing with the soil disposal problem. An early March editorial ignored the Montclair matter completely and complimented the DEP's "admirable initiatives to fight the radon hazard."[46] Later in the month, an editorial noted, without assigning blame, that the department had been "traumatized" by its inability to solve the soil disposal problem.[47]

In May, while criticizing the suggestion that the Montclair soil be stored at a public park, the paper empathized with the DEP's dilemma in overcoming the "hysterical local opposition" that had greeted more "sensible" plans.[48] But early in June, it flatly condemned the DEP's "bungling" efforts to find a location for the Montclair soil.[49] By the end of the month, it again seemed to place the greater responsibility on citizens who reacted "hysterically" to the DEP's plans.[50]

In another swing of the pendulum, two editorials in July referred to the DEP's various soil disposal efforts as "comic opera" and "mind-boggling."[51] Yet in September, *The Record* again ignored the DEP's ongoing soil policy blunders. Instead, it offered unqualified backing to the department's policy of "urging homeowners…to have their houses tested as quickly as possible."[52]

One year later, in October 1988, *The Record* ran an unusual editorial. It acknowledged that "some scientists are having second thoughts about the wisdom of widespread testing." It even concurred that "in some areas, the need for radon testing is debatable." But not in New Jersey. The state's Department of Environmental Protection had indicated that soil in New Jersey might yield more radon than elsewhere, the newspaper reported. Thus, "in New Jersey, there's no need for debate at all."[53]

Following its own advice, *The Record's* editorial page never again expressed doubts about state or federal radon policies. During the next 4 years, it published only one editorial concerning radon. In May 1990, it endorsed New Jersey Senator Frank Lautenberg's proposed bill to require testing for radon in schools throughout the country.[54] The paper seemed to feel that since the wisdom of state and federal policy was now incontestable, at least in the minds of its editorial board, there was no need to write further about the matter.

The Differing Emphases of *The Star-Ledger* and *The Record*

In 1992, I spoke with reporters who had covered the radon issue for the two New Jersey newspapers. Several suggested that *The Star-Ledger* was close to state officials, and the paper generally reported state administration activities favorably. A conversation with Gordon Bishop, a 33-year veteran reporter with the paper, buttressed this impression. By his own count, he had written more than 100 articles on radon.[55] "I broke the story in New Jersey," he explains. "In 1985, Gerry Nicholls came to me and said, 'we've got the numbers.'" Bishop was referring to Gerald Nicholls, the DEP's assistant director for radiation protection programs. Nicholls had presented him with findings that elevated radon levels were present throughout the state. "I'm the one who sounded the alarm," Bishop reiterates with pride.

Bishop obviously was a sympathetic purveyor of the official message, and it was no surprise that he was invited by Senator Frank Lautenberg to travel with him on a radon fact-finding mission the following year. "I was the only reporter who went with his group to Sweden in 1986," Bishop says. *The Star-Ledger's* uncritical assessments of state and federal radon policies evidently place Bishop in good stead with government officials. When I asked if he had anything to do with the paper's editorial position, he responds "I wrote some of the radon editorials."

Bishop doubtless believes that residential radon is as hazardous as the officials have said it is. He feels that his articles reflected a responsibility to educate the public about the hazard and not to introduce ideas that would compete with the official view. I asked why the number of radon articles declined sharply after the mid-1980s. His first reaction was *not* that the issue was no longer news. Rather, he responded as if he had completed the job of informing his readers, and now it was up to them to act. "I can't put a bullet to their heads to do something," he says.

In his recent book, *Health in the Headlines*, Stephen Klaidman also accepts the EPA's presumptions about radon. Formerly a journalist for *The New York Times* and *The Washington Post*, he believes the press should more vigorously be informing people about the need to test and remediate. Indeed, he thinks the press's responsibility to educate differs little from its obligation to inform.

Klaidman's message is ironic. Elsewhere in his book he warns that as a general principle, reporters should "not uncritically accept assertions, even if they are attributed to an apparently reputable scientist."[56] But in his chapter on radon, based on the words of one official from the EPA and another from the DEP, he says that "insufficient attention is being paid to a significant cause of lung cancer."[57]

In contrast, Peter Overby, who worked for *The Record* from 1983 to 1990, seems more sensitive to the press's lack of skepticism on the issue. A newly hired reporter at the end of 1983 when the Montclair radon story surfaced, he became the paper's "radon guy."[58] While he and others on *The Record's* news staff largely reported the official line, their articles also occasionally indicated confusion and ineptness by government officials. In retrospect, Overby thinks that "the critics of EPA were probably not given adequate news coverage." But he recalls how elusive the matter seemed. "It's such a tough issue to get hold of—none of us in the newsroom had a broad sense of what ought to be done."

Another *Record* reporter, who asked not to be identified, feels that the DEP and EPA sometimes used the media for their own public relations. They would call press conferences to announce ostensibly important findings—that the radon threat seemed worse than before or that they had new solutions. But the information often seemed thin, the reporter says. With a tinge of embarrassment, the reporter continues. "We usually just reported what they said."

Based on the evidence, although *The Record's* support of official policy was more conditional than that of *The Star-Ledger*, *The Record* also fell short of unbridled inquiry. In concluding this assessment of the press's role in the radon issue, the section below suggests how a particular incident might have been better handled. It related to a report by state officials presented at a press conference in 1989.

The New Jersey Case-Control Study

A poignant story that should have been written, but was not by any of the papers, involved a report released in August 1989 by the state Department of Health. The report, titled "A Case-Control Study of Radon and Lung Cancer among New Jersey Women," offered the results of a 4-year epidemiological study that sought a correlation between residential radon levels and lung cancer.[59] In conjunction with the release of the report, New Jersey's Health Commissioner Molly Coye described its major alleged finding at a press conference: Radon concentrations below the 4-picocurie action level were causing lung cancer.[60]

Rather than sift through the findings, or ask qualified scientists to do so, *The Record* and *The Star-Ledger* simply reported Commissioner Coye's observations. *The New York Times* did not mention the study at all. Questions about assumptions that underlay radon policies should have been apparent from a superficial examination of the study. But no reporter took on the task. Instead, the commissioner's description of the findings at a press conference became the unchallenged message. "Most of the people who lose their lives are exposed to relatively low levels of radon," said Coye. "There is danger even at the lowest levels."[61]

What did the study actually show, and what should reporters have been asking? The investigation was an extension of a statewide survey of all female New Jersey residents who had been diagnosed as having primary lung cancer from August 1982 through September 1983. About 1,000 cases comprised the original study. These were matched with an equivalent number of controls—that is, women who did not have lung cancer but otherwise shared characteristics with those who did, such as age and race.[62]

In 1985, the state Department of Health began an effort to assess the relationship of radon exposure to the lung cancer cases. The extended study sought to identify the New Jersey residence in which each case and control subject lived longest. Criteria included the requirement that the subject must have lived at the location for at least 10 years during a period between 10 and 30 years earlier. (The 10-year period immediately prior to diagnosis was assumed to be a latency period and was discounted.)

When investigators were able to locate the residences and obtain cooperation of the current inhabitants, they sought year-long radon readings. The study indicates that the residences of 433 cases and 402 controls were successfully tested, principally with alpha track detectors that were retrieved 1 year after remaining in place.[63]

A three-page executive summary of the report states that the relative risk estimate was not statistically significant. But it tries to justify previous assumptions about risk from low-level radon by underscoring that "the *trend* (original emphasis) for increasing risk with increasing radon exposure was statistically significant." The summary then mentions that smoking is the major cause of lung cancer and that "some of the study's results must be interpreted cautiously because of the small number of subjects in the highest radon category." But in bold print, the summary stresses that

> [n]evertheless, the study suggests that the findings of radon-related lung cancer in miners can be applied to the residential setting. Excess radon exposures typical of homes may increase risk of lung cancer; extremely high residential exposures would be associated with very serious lung cancer risks.[64]

A reading of the study provides scarce warrant for the emphasis the authors place on this passage. Nowhere in the summary is there noted the contradictory findings that appear in the body or appendix of the report. For example, when the women who comprised the cases and controls in the original study were assessed according to their counties of residence at the time of diagnosis, the trend was confounding. In counties with the highest radon levels, the proportionate number of cases and controls was about equal (15.1 percent and 14.7 percent). But in counties with the lowest radon levels there actually were more lung cancer cases than controls (30.2 percent compared to 26.6 percent).[65]

Similarly, the study showed curious findings concerning the risk of lung cancer from radon exposure among smokers. The estimated risk increased with radon exposure for light smokers (less than 15 cigarettes per day) and to a lesser extent for moderate smokers (15 to 24 cigarettes per day). But "for lifetime nonsmokers, the pattern was inconsistent," and "paradoxically, the heavy smokers showed a pattern of decreasing odds ratios with increasing radon exposure."[66] Nowhere in the executive summary is there a hint of the finding that with increasing radon exposure, heavy smokers seemed to fare better than light or moderate smokers.

The most damaging of all inconsistencies related to the year-long tests in the homes of the subjects. According to the Department of Environmental Protection, about one-third of all homes in New Jersey have radon levels that exceed 4 picocuries per liter.[67] This means that in a random sample of any 433 New Jersey residents, some 144 should be living in homes with radon concentrations in excess of 4 picocuries. The 433 cases in this study represented an unbiased selection from *all* female lung cancer cases reported during a single period. Therefore, even more than 144 homes might have been expected to show higher than 4-picocurie levels. Yet the number of such homes came to a grand total of six. The number of homes with higher than 4-picocurie readings among all 835 cases and controls totaled eight—less than 1 percent.[68]

In the report's appendix, the authors allude to the discrepancy with the DEP's projections and try to explain it largely as a function of differing sample techniques. The DEP's study was a geographically stratified sample, and the case-control study was a population-based sample. Moreover, the authors presume that the homes of the lung cancer victims were older than the DEP's sample. If so, they conjecture, this also might affect radon readings.[69]

Based on the evidence provided, however, it takes a leap of faith to presume that the discrepancies can be as easily reconciled as the DOH spokespersons suggest. What is clear is that the authors of the study were not interested in focusing on these uncertainties. None was cited in the executive summary, nor were they discussed in press statements or during press appearances by DOH representatives; nor, apparently, did any newspaper reporter ask about them.

A fair appraisal of the study would have challenged Deputy Health Commissioner Thomas Burke's claim that this was the first study that documented risk from low-level radon exposure in homes.[70] The findings should have been reported as ambiguous at best. They raised more questions than they answered.

From the standpoint of public policy, the government officials who presented the report's ostensible findings made extravagant and unwarranted claims. A reading of the full report by a scientifically literate news reporter should have yielded a searching story. It should have led to challenges of the pronouncements by state officials.

Few reporters possess the minimal scientific background to enable them confidently to have mounted the inquiry. Rather than accept this as an excuse, however, it speaks to the need for strengthening scientific literacy among investigative journalists. As June Goodfield rhetorically asks: "[I]f it is not the media's job to promote or assist the public with scientific analysis and debate, then whose job is it?"[71]

In sum, *The New York Times*, *The Star-Ledger*, and *The Record* printed more than 1,000 radon stories in the 1980s. The papers occasionally included news articles that questioned the validity of state and federal policies. But these were infrequent, and critical editorials were rarer still. Far more often, the press echoed dutifully and without analysis the establishment view. In no instance did a newspaper engage in a sustained investigation about the wisdom of the government's radon policies. Most striking of all was the absence of articles on the continuing debate within the scientific community about the presumed hazards of indoor radon.

Notes

1 June Goodfield, *Reflections on Science and the Media* (Washington, DC: American Association for the Advancement of Science, 1983), 94.

2 Rae Goodell, "Problems with the Press: Who's Responsible?" in *Science Off the Pedestal*, eds., Daryl E. Chubin and Ellen W. Chu (Belmont, CA: Wadsworth Publishing Co., 1989), 32.

3 Dorothy Nelkin, *Selling Science: How the Press Covers Science and Technology* (New York: W.H. Freeman and Co., 1987), 174–75.

4 Stephen Klaidman, *Health in the Headlines* (New York: Oxford University Press, 1991), 17.

5 Quoted in ibid., 74.

6 Peter N. Spotts, "'Big Science' Trend Worries Experts," *Christian Science Monitor*, 4 Sept. 1987, 3.

7 William K. Stevens, "Researchers Find Acid Rain Imperils Forests over Time," *New York Times*, 31 Dec. 1989, 1.

8 This is apparent from a review of articles about radon listed in the National Newspaper Index. Television and radio stories about radon, though fewer in number than those in the press, tended also to reflect the official wisdom. One exception to the general media approach could be found in the *Washington Times*, particularly by its columnist Warren Brooks. Brooks often criticized radon and other EPA policies. Many people suspected the newspaper's credibility, however, because it is owned by the Reverend Sun Myung Moon's Unification Church.

9 Coverage of indoor radon in magazines and television news shows between 1984 and 1986 suggests similar patterns to those in newspapers during that period. Allan Mazur, "Putting Radon on the Public's Risk Agenda," *Science, Technology, and Human Values*, Vol. 12, Nos. 60 and 61 (Summer/Fall 1987), 87.

10 Philip Shabecoff, "A Major Radon Peril Is Declared by U.S. in Call for Tests," *New York Times*, 13 Sept. 1988, A-1.

11 "Radon in Jersey Poses Grave Risk," ibid., 1 Oct. 1989, 41.

12 Elizabeth Auster, "EPA Heats up Radon Warning," *Record*, 19 Oct. 1989, A-16.

13 "Fight Radon; Stop Smoking," *New York Times*, 15 Sept. 1988, A-34.

14 Leonard A. Cole, "Radon Scare—Where's the Proof?" *New York Times*, 6 Oct. 1988, A-31.

15 Tom Johnson, "Governor Speeds Cleanup of Radon in Essex Homes," *Star-Ledger*, 2 Dec. 1983, 1.

16 "DEP Officials Assure Swift Work on Glen Ridge Radon Cleanup," ibid., 16 Feb. 1984, 73.

17 Kevin Dilworth, "State Officials Remain Undecided on Radon Cleanup Date, Disposal," ibid., 4 Oct. 1984, 43.

18 Caryl R. Lucas, "Essex Group Seeks EPA Help on Radium Cleanup," ibid., 27 Aug. 1989, 43.

19 Kevin Coughlin, "'Surprising' Change in DEP Policy on Radon Sows Confusion," ibid., 31 Jan. 1989, 17.

20 Ibid.

21 Tom Johnson, "DEP Finds Radon Most Serious Health Peril," ibid., 29 Sept. 1989, 1.

22 Gordon Bishop, "Swedish Expert Calls Radon 'Totally Solvable,'" ibid., 8 Jan. 1986, 1.

23 J. Scott Orr, "EPA Chief Stresses Radon Peril, Urges Testing by All Homeowners," ibid., 19 Oct. 1989, 19.

24 "Worse to Come?" ibid., 7 Dec. 1983, 22.

25 "Precarious Limbo," ibid., 24 Oct. 1985, 26.

26 "Second Thoughts," ibid., 21 Oct. 1986, 16.

27 "The Radon Dilemma," ibid., 20 May 1987, 22.

28 "Rescue Operation," ibid., 5 Sept. 1987, 22.

29 "Silent Peril," ibid., 18 Sept. 1988, Section 3, 2.

30 Tom Johnson, "State Study Finds No Signs of Cancer 'Epidemic' in 3 Radon-Tainted Towns," ibid., 20.

31 Peter Overby and Jan Barry, "Radiation Report Greeted Angrily," *Record*, 2 Dec. 1983, A-1.

32 Ibid.

33 Bettina Boxall, "Radiation Cleanup Begins Tomorrow," ibid., 4 Dec. 1983, A-1.

34 Bettina Boxall and Peter Overby, "Invisible, Odorless Threat Jars 2 Towns," ibid., 11 Dec. 1983, A-49.

35 Peter Overby, "Few Cancers Found at Radon Sites," ibid., 22 Dec. 1983, C-1.

36 Ibid.

37 "Radium in Soil to Be Excavated," ibid., 18 June 1984, A-3.

38 Kathleen O'Brien, "DEP May Ship Toxic Soil West," ibid., 6 Nov. 1984, C-1.

39 Bettina Boxall, "Radon Study Produces No Solid Answers," ibid., 8 Nov. 1985, B-2.

40 James Dao and Dean Chang, "Many N.J. Homeowners Tune Out Threat of Radon," ibid., A-4.

41 "Radon Home-Test Warning Attacked," ibid., 3 Oct. 1988, A-1.

42 David Blomquist, "N.J. Offers Loans to Get Rid of Radon from Houses," ibid., 24 Dec. 1986, C-1.

43 Elizabeth Auster and Peter Overby, "EPA Finds Radon Peril in Survey of 10 States," ibid., 5 Aug. 1987, A-1.

44 Scott J. Higham, "Radon Report May Aid Push for Stricter Limits," ibid., 9 Nov. 1987, A-6.

45 "Radon: Danger Down Under," ibid., 31 Oct. 1985, A-34.

46 "Radon Apathy," ibid., 8 Mar. 1987, O-2.

47 "The Thorium Standoff," ibid., 20 Mar. 1987, A-28.

48 "Montclair's Radon Rumor," ibid., 13 May 1987, A-22.

49 "Radium: The Hunt Goes On," ibid., 2 June 1987, A-22.

50 "Radon: Ignorance Isn't Bliss," ibid., 30 June 1987, B-10.

51 "A Time to Cooperate," ibid., 14 July 1987, B-10; "Radon Soil: The Beat Goes On," ibid., 28 July 1987, B-14.

52 "Radon: Time to Get Serious," ibid., 18 Sept. 1987, B-10.

53 "Radon: No Debate in N.J." ibid., 4 Oct. 1988, B-10.

54 "Radon Alert for Schools," ibid., 29 May 1990, B-8.

55 Interview, June 22, 1992.

56 Klaidman, 231.

57 Ibid., 74.

58 Interview, June 23, 1992.

59 New Jersey Department of Health, *A Case-Control Study of Radon and Lung Cancer among New Jersey Women*, Technical Report—Phase I (Aug. 1989).

60　Harvey Fisher, "Radon's Danger Is Upgraded, Tiny Amounts Can Kill, Study Finds," *Record*, 23 Aug. 1989, A-1, A-14; James Berzok, "Study Finds 'Low' Radon Exposure Can Still Multiply Lung Cancer Risk," *Star Ledger*, 23 Aug. 1989, 1,15.

61　Ibid.

62　Of 1,306 women with lung cancer who were originally identified, interviews were completed with 994 or their next of kin. Among the 1,499 initial controls, interviews were conducted with 995. *A Case-Control Study*, 9.

63　Ibid., 15.

64　Ibid., iv–v.

65　Ibid., App. B13.

66　Ibid., 18–19.

67　New Jersey Department of Environmental Protection, *Highlights of the Statewide Scientific Study of Radon* (Sept. 1989), 1.

68　*A Case Control Study*, 67.

69　Ibid., App. G2-G3.

70　Fisher, A-14.

71　Goodfield, 13.

CHAPTER 9

Radon Policies in Other Countries: Sweden and Finland

ERNEST LÉTOURNEAU, DIRECTOR OF CANADA'S RADIA-
tion Protection Bureau, showed a world map to an audience of scientists and science writers in 1991. Seven countries were highlighted,
the only ones with proposed or adopted radon standards—Canada,
Finland, Germany, Norway, Sweden, the United Kingdom, and the
United States. "As you see," he joked, "radon is a disease that spreads
from the north."[1]

"The geographic configuration is related to luxury," Létourneau
said, "the luxury of worrying about risk that most countries don't feel
is worth worrying about." He did not elaborate on his observation, but
a picture came to mind of the wealth differential that informs this
luxury. The societies with radon policies are affluent. One imagines
only with a sense of absurdity that countries like Bangladesh or
Ethiopia might be considering radon policies.

Létourneau spoke past wealth, however, to culture and values. "I
talked a lot about this with people from different countries, and there
really is a difference in thinking between peoples." He dismissed the
possibility that variations in radon concentration could account for the
differences. "We know now that in all countries, including countries at
the tropical latitudes, you get indoor radon levels comparable to those
of temperate countries." In his view the Nordic, Anglo-Saxon, and
German societies have more activist environmental values than the
French, Latins, and others.

Some countries have not instituted radon controls because of
doubts about the risk, others because they see radon regulation as an
inefficient way to control lung cancer. The French believe there is no

risk demonstrated in houses, "so they are not going to do anything about it," according to Létourneau. The Swiss attitude, he said, is that the gas has been around forever, so why rush into controls. Like many others, the Swiss seem to be awaiting epidemiological evidence that indoor radon is a hazard before they act.

Létourneau's conjecture about cultural influences is reflected in the responses to policy recommendations by the European Community (EC). In consultation with scientific and technical experts, the EC Commission advised in 1990 that countries adopt a 10-picocurie action level for existing structures and a 5-picocurie design level for new construction.[2] But few EC countries were stirred to action. A year later, none of the 12 member states had proposed or adopted controls except the two that previously had done so, Germany and the United Kingdom.

The notion of culture as a policy explanation seems relevant as well to differences among the societies that have adopted radon standards. An examination of radon policies in Sweden and Finland highlights the point. Both countries had developed information and policies about indoor radon well before the United States did. But like all countries with radon policies, their recommended action levels for mitigation are higher than in the United States. While the U.S. action level is 4 picocuries of radon per liter of air, for existing homes in Sweden it is 10 picocuries, and in Finland 20.

The rationale for establishing these levels, the political and social influences that framed them, and the responses of officials and other interested parties in the matter provide valuable comparative information. They lend perspective to the development of radon policies in the United States. They suggest by example the wisdom of a more temperate approach.

Governments in Sweden and Finland

Although Sweden's head of state is a constitutional monarch and Finland's an elected president, the political systems of the countries are similar. Both are parliamentary democracies. The welfare ethos in both is more entrenched than in the United States, and the governments in each country rule with a strong dab of socialism along with free enterprise.

In Sweden and Finland, the political party, or coalition of parties, that holds a majority of parliamentary seats forms the government. The leader of the majority party, or of the strongest party in the coalition, commonly becomes prime minister. As head of the government, he presides over a cabinet of ministers. As in the case of U.S.

cabinet secretaries who are responsible for particular departments, the ministers head the various ministries—from foreign affairs to finance.

Unlike in the United States, the ministers are elected members of parliament. Their subordinates, however, are permanent civil servants who serve irrespective of which parties control the government. The line of authority in each ministry extends to lower level boards and agencies whose functions relate to the interest of the ministry. An issue may be of overlapping interest to several ministries. In Sweden, Finland, and the United States, radon policy is seen as an appropriate concern of several departments or ministries—those concerned with health, environment, buildings, and more.

In practice, the locus of responsibility for radon policy lies with a particular authority. In Sweden, it is the Radiation Protection Institute; in Finland, the National Board of Welfare and Health; and in the United States, the Environmental Protection Agency. But in the three countries, other agencies also share responsibility for indoor radon policies. In the United States, important roles are played by the Department of Energy and to a lesser extent the departments of Health and Human Services and Housing and Urban Development.

Radon Policy in Sweden

The Swedish Radiation Protection Institute (formerly called the National Institute of Radiation Protection) is housed in a low L-shaped brick structure in Stockholm's Karolinska hospital complex. The institute's location seems slightly incongruous, for it is surrounded by other buildings dedicated unequivocally to medical services such as neonatal care, surgery, and radiology. Yet the location symbolizes a prevailing ambiguity about a major concern of the institute—the health implications of residential radon. Whether or not indoor radon proves to be a serious medical threat eventually will define the appropriateness of the institute's location.

The institute's environmental laboratory is headed by Gun Astri Swedjemark, a physicist who has devoted much of her professional life to radon issues. She has worked on the gas and its radioactive daughters since 1971 and has become Sweden's most respected expert on measurement and mitigation techniques. Plain-spoken and direct, she gracefully hosts a visitor and affirms that indoor radon constitutes a health risk that the Swedish public is not sufficiently concerned about.[3]

At the same time, Swedjemark speaks with satisfaction about Swedish policy, which she has helped shape. Her philosophical discomfort with the U.S. approach comes into focus. She criticizes in particular the U.S. EPA's effort to "scare" the American public into action. Rather, she prefers Sweden's effort to provide information dispassionately.

Sweden is a country of keen sensitivity to safety and environmental issues. It has energetically sought to rid its air and water of pollutants: Stockholm is one of a few major cities in the world where swimming in city waterways is safe. The national government required automobile passengers to wear seat belts before they were mandated in other countries. Fear of radiation and the possibility of nuclear plant accidents have inspired a policy to phase out nuclear power entirely.

In these areas and others, Swedish policies have been more aggressive than those of the United States. Yet in the matter of indoor radon, although Swedish concerns came earlier, the policies have been far more deliberate.

Sweden's approach has been all the more interesting in view of the concentrations of the gas found in homes there. Indoor radon in Sweden averages about 2.7 picocuries per liter of air, compared to 1.3 in the United States. (Countries outside the United States commonly express concentrations of radon and radon daughters in international measuring units of becquerels per cubic meter of air. Consistent with the rest of the book, concentrations are described in this chapter in approximate equivalents of picocuries of radon gas per liter of air.)

Development of Policy

The first study in any country of indoor radon or radon daughter concentrations was conducted in Sweden some 37 years ago. In 1956, Hultqvist investigated homes built before 1946 in four towns in central Sweden. Although his measurement technique involved instantaneous sampling that is not now commonly used, contemporary scientists accept the results as valid.[4]

The study found that average radon concentrations for wooden structures were about 0.4 picocurie per liter of air; for brick structures, 1 picocurie; and for concrete based on alum shale, 3 picocuries.[5] The figures seemed hardly alarming, but Swedish scientists continued their interest in indoor radon and remained in the forefront of residential radon investigation in subsequent years.

A study nearly 30 years later found markedly higher average concentrations. Based on 2-week continuous measurements, levels for wooden structures were 2.7 picocuries; for brick-concrete, 1.9; and for concrete based on alum shale, 6.4. The overall average of 2.7 picocuries was nearly four times higher than Hultqvist's earlier average of 0.7 picocuries. The authors of the more recent study suggested several reason for the difference. One was that in the 1980s, houses had enhanced insulation that reduced air change rates and lowered indoor air pressure compared to the soil air. Another was increased use of alum shale–based building materials since the 1940s.[6] But well before the possibility of higher average concentrations was demonstrated, Sweden was addressing the radon issue.

In 1968, a government agency produced the first informational publication by any government on the presumed risks from radon in indoor air. The brochure largely dealt with radon coming from building materials since ground sources were not yet seen as a problem. The brochure's message amounted to warnings about certain building materials and advice that buildings be properly ventilated.[7]

By this time, the relationship of lung cancer to radon exposure among miners had become universally accepted. After calls for a more vigorous approach by scientists at the Swedish Radiation Protection Institute, action was taken in 1974. That year the government halted production of alum shale–based concrete, which when used as a building material caused elevated radon levels. Two years later it placed into general circulation a brochure on indoor radiation hazards. But not until 1978, when a few locations showed unusually high concentrations, did the radon issue attract widespread attention.[8]

The government then appointed a commission of inquiry, which in 1979 proposed a program for action. The following year, the National Board of Social Welfare and Health accepted the commission's recommendations, which included action levels of 20 picocuries for existing buildings, 10 picocuries for rebuilt structures, and 4 picocuries for new construction. At the same time, research efforts on indoor radon were intensified, and the government initiated a public information campaign. In 1983, the recommended action levels became mandatory limits, and in 1990 the limit for existing housing was lowered to 10 picocuries per liter.[9]

Politics of Accommodation

When asked if the official decision to reduce the Swedish standard to 10 picocuries was influenced by the lower U.S. figure, Swedjemark insisted "not at all." The decision, she said, was based on independent assessments of what seemed practicable. Not every Swedish policy maker thinks the decision to move from the 20-picocurie standard for existing buildings was wise. Wilhelm Tell is an instructive example.

Tell has been closely involved with radon policy since the late 1970s. Before retiring from the government in 1989, he was chief of the Building Techniques Section of the National Board of Physical Planning and Building, which included responsibility for health and safety in Swedish buildings. In the 1980s he worked on developing radon policy with officials from other national agencies including those concerned with environment, health and welfare, and radiation. He continues to work as a consultant on building affairs.

In 1991, Tell coauthored with Swedjemark a 100-page monograph that reviewed the knowledge about indoor radon that had been developed during the previous 10 years. Despite their collaboration on the report and a long-term cooperative working relationship, he shares none of Swedjemark's enthusiasm for the new action level. "I am not sure the reduction from 20 picocuries to 10 was a good idea." Tell, a courtly gentleman of 70, slowly works his thoughts into English. A civil engineer, he wonders about the practicality of the move.

> Now we have so many more homes to deal with. When you have 20 picocuries and above, most of the radon comes from the ground. But between 10 and 20 picocuries, most of the radon comes from the building material. It will cost very much to change the value from 18 picocuries to below 10.[10]

Tell knows the spread of costs well. He says that in most homes where the radon source is in the ground, mitigation by way of suction can be accomplished for a Swedish currency equivalent of $2,000 or $3,000. But in houses where radon is emitted by structural materials, the required ventilation can be far higher—at least $10,000, he estimates. He is less optimistic than Swedjemark about the direction of compliance with Swedish policy.

Despite the availability of government loans to help cover mitigation costs, public response had been weak. Tell says that through 1990, approximately 120,000 houses were estimated to have radon levels above 10 picocuries, but only about 1,700 have been mitigated. The consequence is obvious. At the rate that new buildings are being constructed and older ones rebuilt, attaining indoor radon concentrations throughout the country below 4 picocuries (the standard for newly constructed buildings) would take 100 years.[11]

In contrast to Tell, who questions the wisdom of lowering the action level in existing structures to 10 picocuries, Astrid Mäkitalo wishes it were reduced to 5. She has worked on radon policy since the early 1980s, having been associated with the National Board of Health and Welfare. Her training in biology and chemistry provides the basis of her scientific concerns. She now heads the environmental department for a region of Stockholm.

But much as Mäkitalo would like to see tighter standards, she realizes the government is not likely to oblige. She worked for several years to have the standard lowered from 20 to 10 picocuries, and that was difficult enough, she says. For now, "we have a rather good policy in Sweden." Her larger frustration is with an unresponsive public. Although national authorities have labored to alert the public about radon, "we have not succeeded as much as we wanted to. So I don't know what to do. I guess it is that people don't see it, so they don't believe it is a problem."[12]

Both Tell and Mäkitalo have coauthored papers with Swedjemark, and even with each other.[13] Therein lies a distinction from what might be expected in the United States. The three hold clearly different positions about where Sweden's action level for existing housing should be. Yet by way of their collaborative publications, and in my conversations with them, they operate in a spirit of accommodation. Differences among Swedish influentials on radon are devoid of the emotion and antagonism that sometimes have colored the debate in the United States.

Comparing Swedish and American Attitudes

Sweden's action level of 10 picocuries is two-and-a-half times higher than that of the United States. Its policies have been developed more deliberately, and its informational approaches have been more restrained. Swedjemark compares the Swedish and the U.S. approaches.

> I think there is a difference in the two cultures. The national authorities in Sweden agree among each other that we must only give limits and recommendations that are possible to reach. We should not set limits that are not practical. That means we will never be in a situation like that of the U.S. Environmental Protection Agency, for example, which is not realistic. I think they have gone too far. [She refers in particular to the 1988 U.S. Indoor Radon Abatement Act which, with EPA's encouragement, set a goal of reducing indoor radon levels to those found outdoors.]
>
> I have seen the campaign material by the Ad Council, and I found it terrible. You should not frighten people. In Sweden we want the people to realize that radon may be detrimental to them, and we tell them they can do something about it. But not to frighten people, because they can get sick from fright.

Lynn Hubbard

Lynn Hubbard agrees. A physicist at the Radiation Protection Institute, she is uniquely positioned to compare radon policies in the United States and Sweden. Born and educated in the United States, she worked on radon measurement technology at Princeton University before moving to Sweden to continue similar work. In 1991, after nearly 3 years at the Swedish Institute's environmental laboratory, she feels that "people here are more environmentally minded than in the U.S. ...they care more for the environment here."[14]

She also thinks that Sweden's radon policy is more sensible than the U.S. policy. She judges that Swedish people "think a little more" than Americans before acting. "Not that people are slower or lazy; it's just less neurotic here." She considers Sweden's radon policy "about right." The Swedish policy is more laissez faire than that in the United States. The government has issued booklets that advise citizens about what to do if they have elevated radon concentrations, but they carry none of the skull-and-cross-bones flavor that epitomized the EPA/Ad Council campaign. Hubbard is uncomfortable with the U.S. approach.

> We should give people information in the most intelligent way we know. We should put it on a level so that people who don't have Ph.D.s in physics can understand, and they can then do with it what they want. But to scare people as much as the EPA seems to like to do, it's just a little sad to me.

Hubbard tells a revealing personal story. In 1989, after she and her husband contracted to buy a home, a 24-hour radon measurement indicated a level of 12 picocuries per liter of air. They reconsidered whether they wanted the house. Although Sweden's action level at the time was 20 picocuries, they knew that the standard was expected to be lowered to 10. They remained troubled even after the sellers reduced the price by $3,500. In the end they went through with the purchase, but remained uneasy.

> When I measured the house we bought, it was interesting to me that my husband, who is an indoor air quality engineer though not a radon expert, got real depressed and didn't want to buy the house. I felt depressed too, and wondered what we were going to do. Then I thought, isn't this funny, that here we are so knowledgeable, and we still had this reaction.

When asked how she would have reacted if the government advisory remained 20 picocuries, she said she would have felt better.

> Somehow no matter how knowledgeable we are, we pay attention to the rules, because somebody else made them. Somebody else has set that limit because, hopefully, they have thought a long time about what it means. So perhaps I would have thought a little bit less of the reading had the limit been 20 picocuries and there was no view of it coming down. I think that's natural psychology, even though I know so much about the subject.

Since moving to her home, Hubbard has monitored it continuously. The long-term average radon concentration proved to be far below the 12 picocuries indicated in the first test. Although occasionally exceeding 40 picocuries, the radon concentration has averaged less than 2 picocuries. She and her husband feel comfortable about this, and they have not mitigated. Although the short-term test is common in Sweden, as in the United States, its weakness has been made manifest to Hubbard and her family. "It can be extremely misleading," she says.

Competing Interests and Legal Questions

In discussions with people from the radiation, health and welfare, and real estate communities in Sweden, none of the passion found among comparable U.S. groups was evident. This may well be due to the different cultures of the two countries as Swedjemark speculates. Her observation about interest groups in Sweden on the issue of radon was typical of others who discussed the matter. "There are interests here of different kinds, of course, but not very strong. Some say we should have lower limits, but only a very few. And then there are some who think the authorities say the risks are higher than they really are, but not many."

Unlike in the United States where representatives of real estate interests have made clear their skepticism about radon policy, no concerted opposition has developed in Sweden. Ulf Linden, a Stockholm real estate broker, says radon tests are commonly performed when buyers and sellers want them. But like other real estate people, he does not think the issue very important. "I ask the customers if they smoke. If not, I tell them not to have a test—forget it."

Linden says he is unaware of any real estate transactions that did not take place because of radon. He mentions a recent experience when a house had levels slightly above the new action level of 10 picocuries. "We worked it out—reduced the price a little bit." This sounded like Lynn Hubbard's personal experience and comports with her impression that accommodation is typical.

Radon issues have occasionally led to legal actions, nevertheless. Swedish law holds that a home seller must inform a buyer about serious problems of any sort concerning the house. If a buyer later discovers problems, he may sue the seller. In a few instances, parties have gone to court because of disputes over radon concentrations, but these cases have been very rare.[15]

Two other aspects of Swedish policy differ from the U.S. experience. One involves taxes, the other the responsibility of government officials for their recommendations. Local authorities may reduce real estate taxes on homes whose radon concentration is less than 4 picocuries and raise them on homes that exceed 20 picocuries. While intended as an incentive for homeowners to reduce radon levels, the policy has noticeably cut into the public treasury. It has meant an estimated $35 million in reduced taxes in 1989.[16]

The policy will become increasingly costly in terms of tax abatement as more homeowners seek lower radon concentrations. Indeed, the figure represents a substantial portion of government outlays relative to radon. Annual government expenditures for the totality of radon-related activities, including tax abatements, research, measurement, public information and education, approached $100 million.[17] This does not include costs of remediation in private homes, which are borne by the homeowner.

The other singular aspect of Swedish policy relates to the fact that Swedish public officials are answerable for their advice in a way that U.S. officials are not. A national law holds that officials may be sued by members of the public if a company they recommend fails to perform as expected. Astrid Mäkitalo says with a wry smile, "if I recommend a company to perform radon mitigation, and the company does not do proper work, I can be responsible for the cost to the person whom I made the recommendation to."[18]

Agreeing that the responsibility can be somewhat unnerving, Mäkitalo says she protects herself by providing the names of several mitigation companies to inquiring citizens. "I always tell the homeowner what they have to demand from the company. If radon levels are not brought down as agreed upon, then the company must work to get the level lower at no added cost. This should be written in a contract." She has never been sued personally, although she knows another official who has. How was the case resolved? Characteristically for Sweden, the parties compromised. "They made a deal."[19]

Policy Conundrums

While national authorities have developed the framework for radon policies, the responsibility for applying the limits and recommendations lies with local authorities. These authorities are supposed to provide and pay for testing. The homeowner is responsible for mitigation, however, although loans for 50 percent of costs up to $5,000 are available from the government. But at the end of the 1980s, only half of Sweden's 278 municipalities were even taking radon into account when issuing building permits or considering plans for new buildings.[20]

Thus, whatever the differences between the Swedish and U.S. approaches to radon, the two countries face a dilemma common to radon policy authorities everywhere. If radon is perceived to be a serious threat, as Swedish authorities think it is, the country's record of gaining public cooperation has been no more impressive than that of the United States. By 1991, after 10 years of official urging, only 150,000 homes had been tested out of Sweden's 3.8 million (about 2 million are free-standing houses, and the rest attached dwellings and apartments).

Moreover, testing is commonly of the short-term variety that involves a 3-day screening. As discussed earlier and highlighted again by Lynn Hubbard in this chapter, such measurements can be misleading. An added complication in Sweden is that eight different measurement techniques are approved for use. The large number not only may confuse the consumer or local authority who must choose among them, but is inimical to efforts toward consistency in measurement approaches.

Swedish radiation officials hope that by the year 2000, all homes with radon concentrations over 10 picocuries might be found and remediated.[21] But achievement will be difficult. It involves finding an estimated 120,000 homes (3 percent of Sweden's housing) that are presumed to exist throughout the country. An enthusiasm not yet evidenced by local authorities would be necessary to reach this goal. As Tell and Swedjemark recognize: "The scope of the information problem is illustrated by the fact that around 1,000 official persons in 284 municipalities, around 100,000 building professionals, and a large proportion of the more than 8 million population, must be reached."[22]

Formulation and implementation of radon policy in Sweden has been deliberate, but directed. There is a sense about where it is going even if aspects of the timetable seem overly ambitious. Tell and Swedjemark note that because of official efforts and media attention, "it has been possible to develop the radon issue from practically nothing in 1979 to a position in 1989 where the chances of technically and organizationally combating the radon problem are very good."[23]

Radon Policy in Finland

Radon policy in Finland, though also devoid of passion, is more ambiguous than in Sweden. Policy makers in Finland seem undisturbed about the ambiguity.

The Finnish Center for Radiation and Nuclear Safety, known by its Finnish-language acronym as STUK, is similar in scope and authority to Sweden's Radiation Protection Institute. Like its Swedish counterpart, STUK is responsible for research and technical information concerning radiation from all sources, man-made and natural. Its purview includes radon investigation—determining where in Finland the gas is concentrated, how best to measure indoor levels, how to mitigate. Representatives of STUK have participated during the past decade with members of other Finnish agencies in recommending regulatory standards. But the center's role in national policy formulation has not been as clearly defined as that of the Swedish institute.

STUK is represented in the National Agency for Welfare and Health, a body that was created in 1991 by combining the two previously separate boards for welfare and health. The agency is charged with recommending and interpreting radon policies. Before the merger this had been a responsibility only of the National Board of Health in cooperation with STUK.

The building that contains STUK's environmental laboratories is strikingly similar in configuration to Sweden's Radiation Protection Institute. Both are housed in low-flung L-shaped structures, although the Swedish building is covered with brick, the Finnish with concrete and glass. Moreover, the Finnish environmental laboratories occupy only part of the building. The largest portion belongs to offices for adult education. Indeed, the most prominent identifying sign on the building translates as "Professional Education Institute."

Symbolism about the health implications of radon attaches to the location of the Swedish institute in a medical complex. But a different symbolism bears on the Finnish laboratory's presence in a facility devoted to education. Radon officials around the world often regard public apathy about the gas a consequence of inadequate information and education programs. In Finland, however, apathy is explained more as a reflection of national culture.

Olli Castrén, head of STUK's laboratory for natural radiation, seems ambivalent about whether Finland's policy should be more vigorous. He is reluctant to comment directly on the matter—"which is for another agency to do." Tousle-haired and careful with words, he concentrates on the scientific accomplishments of his laboratory. He has been at the Finnish center for 30 years and is the country's foremost authority on radon measurement and mitigation. He explains his interest in radon.

> I have chosen to study natural radiation because the radiation doses are highest there, and to me that makes it the most interesting radiation subject. If radiation protection is important anywhere, it should be natural radiation that should be investigated. We should know about it; that is the minimum. What people are obliged to do is a different matter.[24]

I ask him if he wishes people would do more by way of measuring and mitigating.

Maybe, I don't know. We have our system in Finland, and it's the best system for us. The main thing is that we have informed people, and they are able to make decisions themselves. We have not kept any secrets from them.

Castrén's apparent disinterest seemed typical. Among a score of thoughtful Finns with whom I spoke, including scientists, policy makers, teachers, real estate agents, and homeowners, none expressed serious concern.

Development of Policies and Paradox

Like Sweden, Finland is an affluent country, technologically advanced, and sensitive to environmental issues. But its restrained attitude about radon is even more pronounced than Sweden's. Although radon has been found in certain areas at elevated concentrations, the country's national culture frames its policy of restraint.

After finding unusually high levels of the gas in a few locations in 1981, STUK proposed that an action level of 20 picocuries be set for existing homes, and a 5-picocurie maximum be designed into new homes. The levels were made official in a 1986 directive by the National Board of Health.[25] In 1991 these limits were still in effect, although some officials anticipated that the figure for existing homes might eventually be reduced to 10, following Sweden's example.

The Finnish experience is ironic, because the country may have the best technical program for measurement in the world. Since 1980, virtually all testing has been conducted by highly competent scientists and technicians from a single agency, STUK. Moreover, only one kind of measuring device has been used—alpha track detectors, which are placed in a home for 2 months of continuous monitoring. Since 1980, more than 35,000 homes have been tested in this manner.

In other countries, various government agencies and private companies take radon measurements using different devices and techniques. In Sweden, dozens of commercial firms have performed radon measurements, and in the United States, more than a thousand. The people who take the measurements may have different approaches to placing and locating measuring devices, and tests are commonly conducted for only a few days. Inaccuracies thus become possible for several reasons: varied skills among testing personnel, inconsistency in measurement technologies, and misleading results from short-term measurements that do not represent long-term averages. None of this happens in Finland.

Yet while testing in Finland has been of the highest caliber, the

number of Finnish homes mitigated during the past 10 years is remarkably low: around 150.[26] Among Finland's 2 million dwellings, as
many as 30,000 are estimated to have radon concentrations higher
than 20 picocuries, and 50,000 between 10 and 20 picocuries.[27] In
view of how few homes have been mitigated under the current policy,
lowering the action level to 10 picocuries appears ludicrous. Yet that is
what some in the Finnish government expect to happen.

Risto Aurola and the Influence of National Character

Risto Aurola began working with the National Board of Health in
1978. Soon after, he was assigned responsibility in the radon area
when high levels were found in a few houses. One of his jobs was to
help write the board's 1986 radon directives. Beside confirming the
action level for existing housing at 20 picocuries, the directives encouraged local authorities to test homes in cooperation with STUK.
He now works with the board's successor, the National Agency for
Welfare and Health, where he oversees the implementation of the
radon program by local authorities.

Aurola received a masters degree in public health engineering from
the London Imperial College in 1972. Afterward, he held United Nations
and teaching posts in European countries outside Finland. He draws on
his international experience as he describes the Finnish approach to
public issues—somewhere between German efficiency and Russian inef-
ficiency, he says. Blond and lean, he ponders over his country's radon
policy. "Taking into account the resources that we have, I think the policy
is more or less adequate. But like many areas of official business in this
country it lacks a bit of muscle."[28]

He relates the lack of strong feeling about radon to national
culture.

> There are no groups here lobbying about radon as you have in the
> United States. No environmental groups or anybody else are lobby-
> ing for stronger or weaker policies. That is typically Finnish. There
> are some people who are keen on environmental issues and are
> active on other matters. But the radon issue, I think, relates some-
> how to the Finnish people's nature. Finnish people are a little bit
> sluggish. It takes a long time before they react.

The notion of national character surfaces repeatedly in conversations with other experts, although not all describe it the same way.
Castrén notes that Finns seem to take the radon issue less seriously
than Swedes because "in our country the people want to see the effect
themselves." Eeva Ruosteenoja, who in 1991 completed an epidemiological investigation on radon and lung cancer, explains Finnish apa-

thy as a matter of trust. "Finnish people are not very hysterical. They trust authority, and so they assume there is no big problem."[29]

If Ruosteenoja is correct, a greater public response might be expected if the government made the issue seem more important. But the authorities themselves seem apathetic about the subject, and some are plainly skeptical. Ruosteenoja believes that lowering the action level would be a mistake. Her recently completed study, discussed below, found no significant relationship between radon levels and the incidence of lung cancer. The results help shape her current attitude. "I do not support lowering the limit from 20 to 10 picocuries. The costs will be so much higher. I would put the level even higher, maybe to 25 picocuries."

Aurola, who affirms that "no extra radiation is good for people," nevertheless also doubts the wisdom of lowering the levels. Although the existing levels were officially established in 1986 by the National Board of Health, a new radiation law gives the responsibility of action-level policy to the Ministry of Health and Social Affairs. Changes in policy would have to be directed by that body, to which his agency and STUK are subordinate.

"Now it is up to the ministry," Aurola says, and he thinks "the decision will be more politically based than technical." The pressure for change has increased, he said, because Sweden has lowered its limits "and we often follow Sweden." The European Community (EC) has also recommended a 10-picocurie limit for existing houses. Although Finland is not a member of the EC, "in many cases we try to follow its directives." (Ironically, several members of the European Community, including France and Italy, have no formal radon policies at all.)

Nevertheless, Aurola remains personally skeptical about the wisdom of the proposed policy change.

> I would first like to see whether our technical measures are effective or not. What are the overall costs of reducing radon concentrations? What are the results of measures which have been taken? I would also like to see whether epidemiological studies will indicate what the real meaning of all this is. What is the effectiveness? That is the key to the radon issue that is still missing. What is the usefulness of the present system?

Seeking Linkage Between Indoor Radon and Cancer

Prior to 1991, the only studies in Finland that sought a relationship between concentrations of radon and lung cancer had been reported by Castrén. In 1981, when indoor radon was first suspected as an important cause of lung cancer, an "emergency survey" was conducted in an area with the highest lung cancer statistics. The findings were "very comforting," he wrote. "Only one or two houses out of 138 slightly exceeded the preliminary action level of 20 picocuries per liter of air." Moreover, in areas of the country where more than 10 percent of homes were found to exceed the action level, the lung cancer incidence was "very low."[30]

Measurements throughout the country revealed a pattern of radon concentrations in certain geographical areas. Radon was more abundant, for example, above ground areas called eskers, which were formed by gravel deposits by glacial streams 10,000 years ago. From this information, Castrén and his associates at STUK drew a radon map of Finland, which pinpointed locations of high concentration.

When seeking correlations with lung cancer, however, they found nothing significant. They reported "some resemblance" between lung cancer incidence and radon exposure among rural women in one section of the country. But this was not statistically significant. Smoking seemed the principal if not nearly exclusive cause of lung cancer. Their overriding finding was that "the geographical distribution of lung cancer in men has been changing with time and has very little resemblance to the radon map."[31]

The only other Finnish investigation that sought a connection between indoor radon and cancer was conducted by Eeva Ruosteenoja. A physician, she reported in 1991 the results of a 5-year epidemiological study. Conducted under the auspices of STUK and the Finnish Cancer Registry, she compared 238 men who were diagnosed with lung cancer between 1980 and 1985 and 434 controls. Radon measurements were sought in all dwellings they occupied between 1950 and 1975. In some instances measurements could not be attained and had to be estimated, but for most—164 cases and 334 controls—they were available. While allowing for a "range of uncertainty," Ruosteenoja concluded that the "findings suggest that the risk caused by indoor radon exposure is lower than would be expected on the basis of previous studies among underground miners."[32]

Another of Ruosteenoja's findings was that among the 238 cases, only 4 were nonsmokers, and the types of lung cancers they manifested were not commonly associated either with radon or smoking (usually epidermoid or small cell carcinomas). One had adenocarcinoma, a second had large cell carcinoma, a third was listed as "other" carcinoma, and a fourth as unknown. She found hints that in high radon areas the risk of lung cancer increased slightly among heavy smokers, though not significantly, and not at all among light or nonsmokers.[33]

Castrén is cautious about Ruosteenoja's study. He focuses on its allegedly weak statistical power rather than its findings. He had previously criticized the press for reporting in 1987 that preliminary indications from the study offered "proof that radon is quite harmless and does not cause lung cancer."[34] The study itself makes no such claim.

In any case, the final report was endorsed by Finnish epidemiological experts as a solid and valuable contribution. The 110-page report was an academic dissertation for the medical faculty of the University of Tampere. Indeed, Ruosteenoja's failure to find a correlation between indoor radon and lung cancer was consistent with Castrén's own earlier reports.

In conversation, Ruosteenoja readily acknowledges that her study is not the last word on the subject. But she says directly what she had implied in her report. "I am not so much worried about radon. Mainly people should stop smoking."[35]

Confusion of Authority

Two national agencies with principal responsibilities in radon matters are the Finnish Center for Radiation and Nuclear Safety (STUK) and the National Agency for Welfare and Health. Both are in the line of authority under the Ministry of Health and Social Affairs. In addition, the Ministry of Environment, particularly the section concerned about building safety, and the Ministry of Interior, which oversees technical measures in the planning of housing, have some interests in radon policy.

Whereas in the United States the EPA is the regulatory body for indoor radon, in Finland the Ministry of Environment has only a peripheral responsibility in the matter. "We have no active role in radiation protection," says Lauri Tarasti, the highest civil servant in the ministry. As Secretary General of the ministry since its establishment in 1983, he has served under ministers who have come and gone according to political fortune.[36]

Tarasti, a highly respected official, is as generally informed about environmental policies as any person in the country. Yet during an interview he apologizes for his lack of knowledge about radon policy. He acknowledges with an embarrassed smile that he did not have the slightest idea what his country's action level for radon was, either current or contemplated.

Impeccably dressed, with hair combed precisely behind his high forehead, he laments the confusion about the lines of authority concerning radon. "From an administrative point of view, on the matter of radiation we are not well organized. Radiation problems are divided among three ministries." He mentions the Ministry of Trade and Industry, which is concerned with the nuclear power industry, the Ministry of Health and Social Affairs, and his own Ministry of Environment.

The Ministry of Environment has only a diffuse supervisory role concerning indoor natural radiation. "The problem for us is that the specialist center, STUK, works under the Ministry of Health and Social Affairs, and we have nothing formally to do with them." Tarasti expresses frustration with the structure.

In most European countries, he says, radiation protection is a responsibility of an environment ministry. "Here, people on the street believe we have it, and sometimes that is a problem." His ministry has influence insofar as it is involved with overseeing rules that pertain to planning and building. These largely relate to locations and building materials. But the crux of the development of technical information has come largely from STUK in cooperation with local authorities.

Risto Aurola of the National Agency for Welfare and Health says that by law a dwelling must be "healthy." The country's 250 local health boards are required to act in compliance, for example, by taking measurements of radon or other presumed contaminants, if contamination is suspected. In the case of radon, if local authorities have been advised by STUK or have other reasons to suspect that certain areas have high indoor concentrations, they are obliged to sponsor and pay for testing.

Individual home dwellers may seek testing on their own, but then are personally responsible for payment. The cost for testing is about $200 and is performed as described above, under the auspices of STUK. As Castrén has noted, STUK's capacity is underutilized. "We are able to measure 20,000 houses annually, but there are not that many who buy our services."

Thus in Finland, as in countries with more aggressive radon programs, only a small fraction of homes have been measured—some 35,000 in Finland out of 2 million by 1991. But while local authorities may in many cases pay for testing, the cost of mitigation lies with the home dweller. There are only two or three mitigation companies in all of Finland, in part because the demand for services is so low. The cost for mitigation is in the $2,000 to $3,000 range, though for difficult homes the cost could exceed $10,000, several authorities said.

The only financial assistance to homeowners who wish to mitigate derives from a 1987 government decision to provide help "for repairs which are necessary to remove sanitary defects from houses." This includes radon, but the allowable amount is 20 percent of approved costs, with a ceiling of about $1,800. Few people have used this option. As noted, only 150 homes have been mitigated in 10 years.

Quite clearly, neither bureaucrats nor politicians have tried to drive the radon issue in Finland.

Interests

In the United States the two most visible commercial interest groups on the issue of radon are the radon technology companies that test and mitigate and the real estate industry. Representatives of each have lobbied and testified, with radon companies favoring tighter standards and more aggressive policies. Real estate spokespersons, in contrast, frequently contend that the evidence of health damage from indoor radon is not sufficient to warrant more rigorous policies.

In Finland, none of this applies. In the first place, commercial radon companies are virtually nonexistent. Indoor measurements are performed almost exclusively by the government agency, STUK, and there are only two or three companies that perform mitigation. Unlike in Sweden and the United States, Finnish business telephone directories do not list "radon" or "radiation" as a business category. In the United States, short-term radon testing is routinely conducted in several states before a house sale is completed. In Finland, testing in conjunction with house sales is almost never performed.

Kimmo Koho is head of research and development in Helsinki for Finland's largest real estate agency, Huoneistokeskus. His company has branches throughout the country and, in 1991, was involved in 15 percent of all real estate transactions in Finland. This included about 5,000 homes in the Helsinki area. Among these sales, Koho "cannot remember any case where radon came up." He was vaguely aware that local authorities must include consideration of radon in building plans, as they must for all hazards. But he had never encountered the notion that a radon test be part of a real estate transaction.[37]

Koho speculated that in other areas of Finland, where the gas might be viewed as a greater problem, real estate agents could be more aware of it. He personally felt no concern. Radon limits, action levels? He had no notion what the numbers were.

Later, I spoke with two real estate agents in Tampere, a city 100 miles northwest of Helsinki. It is in an area noted by STUK as having higher than average levels. But like Koho, neither agent knew of any housing transactions involving radon measurements among the nearly 500 that were sold through their offices in the previous 12 months.[38]

Jara Ruuska, a city attorney for Tampere, oversees contracts involving building plans and real estate owned by the city. "Once, a few years ago," he recalls, "people became worried when newspapers reported that radiation was found in some places near Tampere. But that has died down, and no one seems concerned now." As with the real estate agents to whom I spoke, he had little sense of Finnish radon policy and had no idea what the action level was.[39]

To almost all Finns, radon is a nonissue. In a country that has perhaps the most consistent technical program for radon measurement and information gathering, this seems a paradox. To say that the Finnish population is unresponsive *because* it has the best information would be misleading. Other reasons are involved. They include cultural characteristics, the sense of hazard saturation that affects populations in other countries as well, and skepticism in the absence of confirmed evidence.

Whatever the causes, however, the reality of the Finnish situation will strike some as sensible: a technical program for obtaining information about radon, an action level that minimizes the number of homes requiring mitigation, and national leadership that chooses to await proof of danger before pressing the public to act. "To the average decision-maker in Finland, the connection between high radon exposures and lung cancer is not yet very evident," Castrén has written. In characteristic Finnish understatement, he continues, "and this may diminish the motivation for remedial action."

What benefit may be gained from this review of the Swedish and Finnish approaches to radon? Their experiences are informative in their own right, but they also provide a line for comparison with the U.S. approach.

To be sure, there are differences between policies in Finland and Sweden: Finnish policies are less palpable and the attitude of decision makers more laid back than in Sweden. But more impressive are their similarities. Both countries are democracies with deep concerns about the environment. The potential dangers from indoor radon are well understood by the science leadership in both societies. Yet radon policies in both are more circumspect than in the United States.

The leadership in the two Nordic countries appears less intent about trying to rouse the public to act than their American counterparts. According to observers cited in this chapter, the cross-Atlantic differences in radon policies substantially derive from differences in national cultures. In the concluding chapter, questions about democracy, risk, and an optimal radon policy for the United States are considered. The influence of culture and values, however, is never far below the surface.

Notes

1   Comments attributed to Létourneau were made at the Science Writers Workshop on "Radon Today: The Science and the Politics," sponsored by the U.S. Department of Energy, Bethesda, MD, Apr. 25–26, 1991.

2   Commission Recommendation of 21 February 1990 on the Protection of the Public Against Indoor Exposure to Radon, *Official Journal of the European Communities*, No. L 80/26 (Mar. 27, 1990).

3   Unless otherwise noted, comments attributed to Gun Astri Swedjemark are from an interview, Aug. 26, 1991.

4   Cited in Gun Astri Swedjemark, Agneta Buren, and Lars Mjones, "Radon Levels in Swedish Homes: A Comparison of the 1980s With the 1950s," *Radon and Its Decay Products*, ed., Philip K. Hopke (Washington, DC: American Chemical Society, 1987), 84.

5   Ibid., 87.

6   This was based on a countrywide study conducted in 1980–1982 of homes built before 1976, ibid., 89–91.

7   Bo Lindell and Sven Lofveberg, "Radioaktivitet Och Byggnadsmaterial" ["Radioactivity in Building Material"], *Byggforskningen Informerar* [*Building-Research Information*] (State Institute for Building Research, Stockholm, Feb. and Mar. 1968).

8   Wilhelm Tell and Gun Astri Swedjemark, *Kunskapstillvaxten Och Dess Effekter Inom Radonomradet* [*The Growth, Dissemination and Implementation of Knowledge Regarding Radon in Buildings*], A Study of Swedish Conditions Between 1979 and 1990 (in Swedish with English summary), Scientific Advisory Board of the Swedish Council for Building Research (Stockholm: Swedish Council for Building Research, 1991), 84.

9   Ibid.; interview with Swedjemark, Aug. 26, 1991; interview with Wilhelm Tell, Aug. 26, 1991.

10   Interview, Aug. 26, 1991.

11   Interview; also see Tell and Swedjemark, 83–87.

12   Interview, Aug. 27, 1991.

13   Gun Astri Swedjemark, Hakan Wahren, Astrid Mäkitalo, and Wilhelm Tell, "Experience From Indoor Radon-Daughter Limitation Schemes in Sweden," *Environment International*, Vol. 15 (1989).

14   Unless otherwise noted, comments attributed to Lynn Hubbard are from an interview, Aug. 26, 1991.

15   Gun Astri Swedjemark, interview, Aug. 26, 1991.

16   Tell and Swedjemark, 59–60.

17   Ibid.

18   Interview, Aug. 27, 1992.

19   Ibid.

20   Gun Astri Swedjemark et al., "Experience from Indoor Radon-Daughter Limitation Schemes in Sweden," 254.

21   Tell and Swedjemark, 83.

22   Ibid., 86.

23   Ibid.

24   Unless otherwise noted, comments attributed to Olli Castrén are from interviews on Aug. 29 and Sept. 4, 1991.

25   Olli Castrén, "Dealing With Radon in Dwellings: The Finnish Experience,"

presented at the Second International Specialty Conference on Indoor Radon, Air Pollution Control Association, NJ, Apr. 6–10, 1987, 2–3.

26 Castrén, interviews.

27 Castrén, "Dealing With Radon in Dwellings: The Finnish Experience," 8.

28 Unless otherwise noted, comments attributed to Risto Aurola are from an interview on Sept. 2, 1991.

29 Unless otherwise noted, comments attributed to Eeva Ruosteenoja are from an interview, Sept. 3, 1991,

30 Castrén, "Dealing With Radon in Dwellings: The Finnish Experience," 2.

31 Ibid., 6.

32 E. Ruosteenoja, *Indoor Radon and Risk of Lung Cancer: An Epidemiological Study in Finland*, Finnish Center for Radiation and Nuclear Safety (Helsinki: Finnish Government Printing Center, 1991), 97–98.

33 Ibid., 69, 95.

34 Olli Castrén, "Radon in Finland," *Finnish Features*, The Ministry for Foreign Affairs (Helsinki: Finnish Government Printing Center, 1990) Document 6.3.7.4, p. 4.

35 Interview, Sept. 3, 1991.

36 Comments attributed to Lauri Tarasti are from an interview, Sept. 6, 1991.

37 Interview, Sept. 5, 1991.

38 Interviews with Ilkka Koiunen and Raia Rikonnen, agents for the Huoneistokeskus real estate company in Tampere, Sept. 6, 1991.

39 Interview, Sept. 6, 1991.

CHAPTER 10

Democracy, Risk, and Reason

IN 1992, MORE THAN 6 YEARS AFTER FEDERAL OFFICIALS first warned that radon in homes was a widespread problem, the American public remained largely unimpressed. To the chagrin of congressional and regulatory officials, few people heeded the Environmental Protection Agency's (EPA) admonition to check their homes for the gas. Only 5 percent of American homes had been tested. Fewer than 3 percent of the homes believed to have radon concentrations above 4 picocuries per liter of air had been mitigated.[1]

While the public remained impassive, federal officials pressed for more aggressive action. Early in the year, the U.S. Senate overwhelmingly passed a bill that would include as a national goal that every home, school, and federal building in the United States be tested for radon.[2] And EPA's new 1992 *Citizen's Guide* affirmed that "no level of radon is safe."[3] While reiterating the 4-picocurie action level put forth in the 1986 *Guide*, the new brochure implicitly encouraged homeowners to act, even if concentrations were lower: "Radon levels less than 4 pCi/L still pose a risk, and in many cases may be reduced."[4] The more rigorous charge to homeowners came after a 3-year advertising blitz about radon failed to rouse the public. An unmistakable dissonance was evident between the citizenry and its political leadership.

The issue speaks to the age-old conundrum about the proper relationship in a democracy between a people and its government. It is also germane to more recent concerns about dealing with health and environmental risks. After discussing these questions in the perspective of the book's presentation, a concluding comment is derived about a reasonable approach to radon policy.

Democracy and Radon Policy

Tension between the freedom of an individual and the needs of society is inevitable to the human condition. Every person's actions at times must be subordinated to the interest of the larger community. The areas of state interest are expansive, and limitations on personal freedom range from mandated traffic behavior to compulsory school attendance. But the ethos of modern democratic society holds that the rights of individuals be maximized and that state impositions be exercised as sparingly as possible.

One of the obligations of the state is to protect its citizens from harm. However, fine lines of guidance do not exist to indicate which harms deserve government protection and by what means. Some cases are beyond dispute. If a country is invaded by outside military forces, it may legitimately draft its citizens into service to resist. If citizens are afflicted with a highly contagious dangerous disease, they may rightfully be segregated to protect others from contracting the disease. But people also engage in hazardous activities that, by general understanding, should not be proscribed by the state. Smoking, eating fatty foods, and riding motorcycles are common examples.

The growth of environmental concerns since the 1960s has broadened the field of inquiry about state intervention. Which environmental risks should be of state interest? What policies should apply? What part should the public have in the process? The degree that governments may intervene in matters of health and environmental protection raises philosophical as well as pragmatic questions.

In 1990, the Environmental Protection Agency produced a list of the most serious environmental problems identified by its Science Advisory Board. Radon was near the top. Yet a survey that year indicated that the public ranked indoor radon 28th in importance out of 29 possible health and environmental concerns. Differences in perception between technical experts and the public extended to other environmental and health issues as well.

EPA spokespersons indicated that the agency would thereafter dedicate more of its resources to the issues identified by the experts. Beside radon they included toxic air pollutants, drinking water contamination, and the greenhouse effect. In consequence, the agency would pay less attention to items highest on the public's list of concerns, such as hazardous waste sites and water pollution from industrial waste.[5]

As Daniel Wartenberg and Caron Chess observe, however, concerns of the public and those of scientists may both be valid. They note that local residents have been sources of information that escaped the attention of scientists and their equipment. Moreover, values vary in different communities that can affect risk evaluation. Finally, "edicts from scientific experts or from regulators routinely infuriate affected citizens who have not taken part in the evaluation."[6] The authors' observations underscore the tangible benefits of including citizens in the formulation of policies. But even if pragmatic consequences were not at issue, public policies should reflect public sensitivities as a matter of democratic principle.

David Bazelon, a former member of the U.S. Court of Appeals, recognized that the public at times may seem irrational in its response to risk. But he dismissed the notion that experts alone should make decisions that affect the general society, no matter how complex the issue. In response to skeptics who think the public incapable of making informed choices, he summoned the words of Thomas Jefferson.

I know no safe depository of the ultimate powers of the society but the people themselves; and if we think them not enlightened enough to exercise their control with a wholesome discretion, the remedy is not to take it from them, but to inform their discretion.[7]

Informing the Public

In searching for optimal environmental policies, some scholars have criticized the democratic model of decision making. Philosopher Douglas MacLean does not think that popular will alone is adequate to make wise choices. Environmental policy decisions must also reflect the application of moral philosophy. "The function of government," he writes, "is not simply to reflect current preferences. Government also has an ennobling and educational role to play, even in a democracy."[8]

MacLean seems to treat preferences and education as unrelated distinctions in a political system. But few people who support a democratic model presume that popular preferences should come from an uninformed citizenry. To the contrary, the responsibility of government includes education, so that citizens may make informed choices.

With respect to providing information about radon, the content and manner of presentation have become issues of contention. In its zeal to convince homeowners that indoor radon can be hazardous and that they should test and mitigate, the EPA has experimented with different informational techniques.[9] Indeed, the EPA's concern about the lack of public response has helped stimulate growth of a new discipline—risk communication. A hardly surprising finding in the risk communication literature is that the way information is presented affects the way it is received.[10]

In principle, educating the public about radon risk is entirely appropriate. But the information, even if hortatory, should not tilt the evidence. Rather, it should frankly acknowledge uncertainties and scrupulously avoid factual distortion. In the words of a National Research Council Report on risk communication, *"deception is never appropriate"* (original emphasis).[11]

As discussed in Chapter 1, however, the EPA/Ad Council campaign has been criticized as alarmist and misleading. Depicting people in their homes turning into skeletons because of exposure to radon, as agency advertisements have done, demeans the education process. In the first place, it suggests imminent catastrophe that, by any measure, is untrue. Second, it glosses over the many assumptions and uncertainties on which the risk estimates are based. Third, it does not respect the right of citizens to receive a full range of information and then to make a choice—and to make the "wrong" choice at that.

Paradoxically, the public has remained largely unresponsive after years of media coverage and EPA-directed information. This has led to calls that the public be required to do what it would not do voluntarily.

Popular Will

Inquiries into democratic political systems commonly focus on structure and process. They include sophisticated analyses of electoral procedures, manner of representation, and governmental and extra-governmental actions. Volumes have been written about how the popular will may be expressed through the maze of political institutions. But the most vivid expression of the will of the people can ignore all the intervening mechanisms and take the form of mass behavior.

Grass-roots discontent has repeatedly forced changes in public policies that otherwise might have remained static. Demonstrations, petitions, and other activist expressions radically altered the nation's policies on civil rights, the Vietnam war, and environmentalism itself. Passive behavior, if less dramatic, can be an equally potent expression of popular desire.

Experts fret that the public has ignored warnings about radon because of ignorance and unwarranted apathy. But after years of publicity and advertising, the citizenry cannot have escaped exposure to the EPA's view of the matter. The magnitude of the EPA/Ad Council campaign alone has been daunting. In 1992, Stephen Page, director of EPA's radon division, said that during the two previous years, radon public service announcements were released to 630 television stations, 3,000 radio stations, and 4,000 newspapers and magazines.[12] At the same time another EPA official indicated that surveys showed that nearly 80 percent of the public had become aware of radon.[13]

Thus, whether or not citizens have absorbed the material in the manner the EPA wished, they have received the information. On the evidence, few have chosen to act. By now the public's behavior may reasonably be interpreted as informed unresponsiveness. Has the behavior been irrational?

In her exploration of risk and rationality, Kristin Shrader-Frechette concludes that "laypersons are often more rational, in their evaluation of societal risks, than either experts or governments appear to have recognized."[14] Anthony Nero, a specialist on indoor radon, thinks the public has an accurate sense of "real risks in various aspects of one's life," and radon is not one of them. "The EPA program," he says, "in decrying perceived public apathy about radon, may be failing to recognize that the public might have a more functionally sound concept of the indoor environment than does the EPA."[15]

The public's lack of response, a most elemental democratic expression, has occurred despite its leaders' admonitions. Moreover, the zeal with which the leaders have pressed the issue has at times challenged democratic precepts.

Openness and Institutional Responses

The degree that decision making about radon has been compatible with democratic values may be gauged from two perspectives: openness and institutional behavior. On both counts, the political system has shown weaknesses.

Where policy choices must be made, openness means placing into the public domain the full range of knowledge, uncertainties, and differing viewpoints about a subject. To the extent that these criteria are circumscribed, so are democratic values. Political and social arrangements in a nation require that a few authorities make day-to-day decisions that affect the health and welfare of the citizenry. But the outlines of policy in a democratic society should reflect the popular will. If the authorities skew or limit the information they provide to the public, the democratic process is subverted.

In the development of radon policy, the principle of openness has at times been threatened. Despite urgings by some to place before the citizenry only the EPA's position, several scientists and non-EPA officials have publicly resisted. But as indicated in this book, they report having to surmount uncomfortable pressure in the process.

A revealing example involves a 1990 article that assessed the radon–lung cancer connection in relation to smoking habits. Using data that provided the basis for EPA's policies, the article concluded that if some 16,000 annual lung cancer deaths were attributable to radon exposure, people who never smoked would account for 500 of the deaths.[16] If the annual total was 7,000, EPA's current low-end estimate, scarcely 200 nonsmokers would be part of the total. The implication of the study was unmistakable: eliminate smoking and almost all the radon problem is solved.

The calculations, though scientifically sound, did not please radon officials at the EPA. The relatively modest number of nonsmokers among presumed radon casualties took the bite out of the agency's customary warning. The EPA has always emphasized the danger to all citizens, smokers or not.

One of the authors of the study, Kevin Teichman, is a mechanical engineer with EPA's Office of Research and Development. (The other author, William Nazaroff, is a professor at the University of California at Berkeley.) By coauthoring the article, Teichman challenged the "everyone-should-be-speaking-with-one-voice" theme that the agency was advocating. According to people familiar with EPA activities, Teichman's article was not received well within the agency, and he has since tried to distance himself from further radon-related work.

Richard Guimond, former director of EPA's radon division, indicated that various views on radon had been debated within the agency during the 1980s. From these debates the agency had developed a policy, and it was now inappropriate "for an individual to scuttle what 90 percent of the agency believes."[17] The formulation has had an effect. Other officials with whom I spoke said they would not publicly express their doubts about current policy because this could jeopardize their careers.

The public is the loser when knowledgeable dissenters are not heard. When this happens, citizens can neither gauge the number of experts who disagree with establishment policy nor know the range of ideas that enter into the disagreement. However well-intentioned, efforts to limit discourse and dissent on the radon issue, as on other issues, should be understood as contrary to democratic values.

The second perspective, institutional responses, has been explored throughout the book. Radon activity by most state governments has been limited, but federal action has been vigorous. Congressional legislation in 1986 and 1988 assigned to the EPA the primary responsibility for developing radon policy. In addition, congressional committees held several hearings on the subject. Interest groups presented a spectrum of views at the hearings, but skeptics from the scientific community appeared infrequently. This was not because they were few in number or unwilling to speak out, as the book has shown. They simply were rarely invited to testify.

As recounted in Chapter 5, members of Congress themselves became spokespersons for aggressive radon policies. By the end of the decade, EPA authorities were expressing great pride in the agency's radon accomplishments. A few senators and representatives had become leapers and pressed for even more aggressive actions than those urged by the EPA. Most, however, quietly accepted the radon issue as framed by the agency. Similarly, newspapers largely accepted the establishment position and reported the issue as defined by the EPA (Chapter 8).

Thus, three central institutions—Congress, EPA, and the press—generously exposed the public to the establishment position and the need for a strong radon policy. Although they provided some opportunities for skeptics to be heard, the institutions offered far more exposure to supporters than critics of the policy.

The weaknesses in the institutional handling of radon policy deserve explication and correction. Their disclosure should serve as a reminder to the nation's leadership about an essential democratic precept: Openness should override even the most sincere conviction that any individual or agency know better what is good for the people than the people know for themselves.

Pattern of Haste

In a lecture after World War II, former diplomat George Kennan advised that Americans should approach world affairs as gardeners and not mechanics.[18] His advice reflected a common perception about American character—that Americans are "preeminently [people] of short views," that they are "compulsive rather than relaxed," and that they "fail to reap the advantages of thoughtful policy-planning."[19] The observations suggest that impatience and the propensity for quick fixes are ingredients of American political culture and that they inhere in the nation's policies.

William Ruckelshaus, a former EPA administrator, spoke to this inclination in the environmental area. His criticism of the first congressional mandate to address air pollution suggests the frustrations that can result from this behavior.

> The Clean Air Act of 1970, for example, gave EPA 90 days from the date of enactment to propose national ambient air standards for the

major pollutants, standards that would fully be protective of the public health, and told us we had five years to attain them. This was done in the face of evidence that the problem in such cities as Los Angeles would take 25 years to solve.

The mandate was one of many that he termed "congressional prescriptions for progress expressed in unattainable goals, technological fixes, and unrealistic deadlines."[20]

The U.S. approach to radon has reflected this pattern. It is patently illustrated in a comparison of radon policy in this country with that in others. As described in Chapter 9, no other country, including several with strong environmental concerns, has radon standards or policies as aggressive as those in the United States.

The common explanation offered by U.S. officials to describe the country's quick response was the discovery in late 1984 of high levels in the Watras home. Inspectors unexpectedly found concentrations of 2,700 picocuries of radon per liter of air there. The surprising discovery rightfully generated concern. But the concern logically should have led to seek out other homes with very high levels. Instead, the Watras affair prompted a national policy that advised mitigation at 4 picocuries, involving an estimated 8 million homes. (In 1992, the EPA lowered the estimated number of homes to 6 million.)

Swedish and Finnish officials ascribed their countries' more relaxed approaches to radon as a reflection of their cultural differences with the United States. They addressed the matter as gardeners, not mechanics.

Risk

An examination of the radon issue in terms of risk analysis reinforces doubts about this nation's approach to indoor radon policy. Heightened concerns in the 1970s and 1980s about environmental hazards and human exposure to toxic substances stimulated interest in the subject of risk. Management of risk was recognized as a political and therefore value-laden process. But risk assessment was another matter. Both aspects of risk analysis are integral to the indoor radon experience.

Risk Assessment

A 1983 National Research Council study embraced the notion that properly performed risk assessment was value-free. "Risk assessment is the use of the factual base to define the health effects of exposure of individuals or populations to hazardous materials and situations."[21]

William Ruckelshaus offered a broader perspective. The former EPA administrator recognized that facts were not the only ingredients involved in the assessment process, but that values were also inherent. Nevertheless, he implied that near objectivity remained possible. "Although we cannot remove values from risk assessment, we can and should keep those values from shifting arbitrarily with the political winds."[22]

The notion that values can be prevented from shifting in a process in which they are inherent is self-contradictory. Which values will be held steady and which allowed to flourish? Constraining a set of values is itself an expression of someone's values.

Risk assessment devoid of bias is a chimera, because it pretends to ignore the inevitable: Gaps in knowledge are filled by human beings based on their best guesses or desires. What Ruckelshaus acknowledged with reluctance, others accepted with unqualified realism. "Far from being objective," wrote Mary Douglas and Aaron Wildavsky about risk analysis, "the figures about probabilities that are put into the calculation reflect the assigner's confidence that the events are likely to occur."[23]

This observation has been amply confirmed in the radon experience. Radon investigators recognize uncertainties about the conditions of the miners from which extrapolations about residential exposures have been made. The range of variables is broad, including smoking habits, breathing rates, work environments, sex, age, and exposure concentrations. In making calculations, a scientist can only estimate the probable effect of each variable. In consequence, reputable investigators have developed differing risk models about indoor radon.

But the imprimatur of the scientists' sponsoring organizations creates an image of vaunted wisdom. Reports issued by the National Academy of Sciences command respect by virtue of the academy's august position in this society. Similarly, studies by the National Council of Radiation Protection and Measurements, an organization prestigious in its own right, are received with high regard. Yet reports from both organizations about low concentrations of radon are based on the *assumption* that *any* level of radon exposure may be harmful.

Even in the light of this common assumption, as discussed in Chapter 2, these two organizations arrived at different risk estimates about indoor radon. Their assessments speak to the essential point: Some elements of risk may be objectively defined, but quantifying the risk from low-level radon is ineluctably a value-laden exercise.

In this regard the radon issue is not unique. Shrader-Frechette notes similar modeling problems with other presumed carcinogens: formaldehyde, ethylene dibromide, dioxin, and methylene chloride.

The EPA estimated, for example, that two people in 10,000 who were exposed to formaldehyde-treated pressed wood over 10 years faced a consequent cancer risk. But experts then produced other conflicting models, as Shrader-Frechette writes, "each with attendant assumptions." Some argued that the EPA's assessment was too high; others, that it was too low.[24]

The complexity of risk assessment has become increasingly recognized as investigators have begun to create separate categories for risks to men, women, and children. In some cases, eating habits are now taken into account in risk assessments of, exposure to certain chemicals—vegetarians as distinct from meat eaters, and people who raise and eat their own vegetables compared to those who do not.[25] While these distinctions have not yet been applied to radon risk models, they reflect an increasing sense of uncertainty in the arena of risk assessment.

Risk Management

Apart from its value-free presumptions about risk assessment, the 1983 National Research Council study recognized that risk management is grounded in values. Yet its description of the process is freighted with so many entry points that choosing a management strategy seems almost arbitrary. "Risk management is the process of weighing policy alternatives and selecting the most appropriate regulatory action, integrating the results of risk assessment with engineering data and with social, economic, and political concerns to reach a decision."[26] The challenge to establish an optimal management policy seems daunting. What lessons from the radon experience are applicable?

If appropriate management means setting a course between critics who call for more action and others who want less, the EPA's radon policies are creditable. In many ways the policies meet these criteria: Some people think the EPA's 4-picocurie action level is too high; others, too low. EPA's call for voluntary testing of every home also takes the middle road; it falls between mandatory testing and testing in limited areas, if at all. But is the middle course the wisest? The answer depends on one's position.

Steven Jellinek, former assistant administrator for pesticides and toxic substances in the EPA, delineates the difference between the roles of the scientist and regulator. Scientists, he says, like to be sure before saying something is true or false. Regulators, however, "do not have the luxury of putting off decisions until certainty arrives." In consequence, "the regulator must step into a murky world of imprecision where scientists fear to tread." The regulator's considerations must include the financial cost of taking or not taking precautionary action, the harm caused by taking or not taking action, and the balancing requirements in federal laws.[27] From the perspective of some regulatory or political officials, EPA's radon advisories may seem optimal.

But from the scientist's viewpoint the issue is quite different. The scientist is usually spared the pressure of opposing groups trying to press their will on his or her conclusions. The scientist is less obliged to leverage the economic cost or legal ramifications of his or her assessments. Thus, the scientist's approach to management policy is likely to be less anticipatory and more reliant on hard evidence. In some areas this caution may delay necessary action, though in others postponement will have proved appropriate. Cases must be judged on their own merits.

In the matter of radon, the crucial scientific determinant should lie in the area of epidemiology. No matter what radiation theory suggests, no matter how sophisticated a risk estimate model may be, the validation of an aggressive management policy should depend on one precept: a demonstrated correlation between lung cancer and residential radon.

According to one investigator, because the number of annual deaths that the EPA presumes comes from radon is so high—between 7,000 and 30,000—if there is an effect, "it should hit you in the eye." Not all scientists agree that the answer is so straightforward. Several stress the need for clear statistical evidence before current risk models can be discounted. The views underscore the uncertainty of the matter and reconfirm that the overriding quest of the scientist should be for epidemiological evidence.

Is the public interest concerning radon better served by the regulator's perspective or the scientist's? The answer cannot be neatly served up. In the same way that matters of risk assessment are hybrids of fact and value, so are environmental and health policies. The value component inevitably coexists with the fact component. But an essential feature of a democracy requires that a citizenry feels that its government properly expresses the will of the people.

The relationship between citizen and leader is ever vulnerable to individual aspirations and misguided assumptions. Even when actions are aimed exclusively at the public good, public trust can be jeopardized. Jeffrey Harris, for example, favors "incremental regulatory steps at early stages of a problem" because, he says, they can be reversed if intervention later proves unwarranted.[28] But he does not address the question of how many times the public will forgive the authors of mistaken policies before losing confidence in their word altogether.

The radon soil issue in Montclair, New Jersey, described in Chapter 6, confirms that public tolerance for official error is limited. Repeated failure by New Jersey environmental authorities to implement policies acceptable to affected citizens led to popular disillusionment. In the end, state officials were entirely excluded from the soil removal program.

If the word and promise of officials prove erroneous, the confidence of citizens in their leaders can only suffer. Recognition of this political truth should signal caution to regulators who would press upon the public a potentially wasteful and expensive policy. To those who worry that hesitancy might result in unwarranted risk and expense, the wisdom of David Bazelon should give comfort: "Delay that is a necessary incidence of calm reflection, full debate, and mature decision more than compensates for the additional costs it imposes."[29]

Radon and Reason

The consequences to individuals from current radon policies can be tangled and expensive. In 1987, Gonzalo Mercado and his wife Maria contracted to purchase a home in Rutherford, New Jersey, for $200,000. Soon after, they learned that a 3-day test in the house showed a radon level of 2.9 picocuries per liter of air. Citing the EPA's "risk evaluation chart," the Mercados said the radon exposure was equivalent to 200 chest x-rays a year. Frightened for themselves and their children, they reneged on the contract.

The homeowner could not find another buyer for months, and then only at a lower price. He went to court, and the judge ruled that since the EPA's proclaimed action level was 4 picocuries, the Mercados were responsible for costs incurred by the other parties. He ordered them to pay the homeowner $16,250 and the broker $6,000.[30]

Other suits have been instituted over radon claims against sellers, architects, and builders. "I would say that what we have seen so far is the tip of the iceberg," says Lawrence Kirsh, a Washington lawyer who advises corporations on indoor air-pollution issues.[31]

Beyond legal action, short-term radon tests have caused innumerable delays or cancellations in real estate transactions. For example, in 1991 Jerry Gersh exercised a contractual option to withdraw from his agreement to buy a home in a Bergen County, New Jersey, community. A 2-day measurement in the basement yielded a reading of 4.7 picocuries. Although assured by a radon-testing company that the level could be lowered, he worried about possible difficulties when trying to resell the house. "When I want to sell, people will want to find a home that never had a radon problem, which is the way I feel now."[32]

Disruptions and personal costs are bound to increase as more homes are tested. Of course, they should be deemed necessary inconveniences if the overall policy saves lives. But if the fuss over indoor radon proves to have been exaggerated, millions of citizens may become embittered.

Does this mean we should shelve a radon policy until there is definitive epidemiological evidence? Not entirely. As well as I can fathom, in balancing cost and risk with common sense, the most rational policy now should favor locating the relatively few homes with the highest radon concentrations and reducing those levels.

Were EPA's action level not 4 picocuries but 20, as is the standard in Canada and Finland, the number of supposedly unsafe U.S. homes would drop dramatically. Instead of the 8 million homes that EPA presumed need remediation, the figure would be about 70,000, according to Nazaroff and Teichman.[33] (The revised EPA estimate of 6 million homes with radon levels above 4 picocuries places fewer than 50,000 with levels above 20 picocuries.[34])

The cost of finding and remediating these homes might reach hundreds of millions of dollars. But the amount is modest compared to the $8 to $20 billion built into present policy. And it is many tens if not hundreds of billions less than would be required to fulfill the legally mandated goal to reduce all indoor radon concentrations to outdoor levels (less than 0.5 picocuries).

Guarding Against Epicycles

Even if the worst suspicions about radon in homes prove correct—that low levels in the home do correlate with cancer—a limited approach, that of the loper, seems the most prudent for now. That is because the risk is associated with long-term exposure. The delay of a few years is unlikely to make much difference in lifetime totals. Completion of several more epidemiological studies is expected in the next few years, and awaiting the results before imposing an expensive and intrusive policy seems a wise course.

What if epidemiological studies continue to show no relationship between cancer and radon in homes? I posed the question to dozens of government officials and scientists. The answers were unsettling. Most doubted that the government would substantially alter the current approach, whatever the findings. Responses ranged from conviction that the studies will surely show a relationship, to belief that if they do not, more studies would be necessary.

There is a hint of Ptolemy in all of this. The second-century astronomer conjured a complex model of epicycles to explain the apparent retrograde motion of the planets. The model was necessary to sustain the prevailing belief that all celestial bodies revolved around the earth. The model carried the wisdom of the time.

When in the 1500s Copernicus and others placed the earth and the other planets in orbit around the sun, their ideas were scorned. The logic of their case—the implicit lack of evidence for an Earth-centered universe—failed to change the belief system. Even Galileo's invention of the telescope in 1610, which showed visual evidence to the contrary, initially failed to sweep away Ptolemy and his epicycles.

Some experts did not believe what they saw through the telescope, and others refused to look. "My dear Kepler," wrote Galileo to his friend about these scientists and philosophers, "what would you say of the learned here, who, replete with the pertinacity of the asp, have steadfastly refused to cast a glance through the telescope? What shall we make of all this? Shall we laugh, or shall we cry?"[35]

Of course, there are important differences between the seventeenth century conflict over heliocentrism and the current radon issue. Yet humans remain no less capable of sustaining beliefs out of habit, wish, or self-interest—irrespective of the evidence or lack of evidence. It is against this unhappy potential that Ptolemy's epicycles should signal caution.

No one has a monopoly on wisdom about what a proper radon policy should be. This essential fact deserves respect from congressional and regulatory leaders who too often speak with certitude about the matter. Since consensus is lacking in the science and health communities, differing viewpoints should be welcomed and broadcast. This is surely the healthiest approach for people in a democracy.

Notes

1   Radon Division, Office of Radiation Programs, U.S. Environmental Protection Agency, *Technical Support Document for the 1992 Citizen's Guide to Radon* (Washington, DC: U.S. Environmental Protection Agency, May 20, 1992), ch. 6, p. 1.

2   "Indoor Radon Abatement Reauthorization Act of 1992," S. 792, passed 82-6, *Congressional Rec.* S2994-98 (Mar. 10, 1992).

3   U.S. Environmental Protection Agency, *A Citizen's Guide to Radon: The Guide to Protecting Yourself and Your Family from Radon*, 2d ed. (Washington, DC: Government Printing Office, May 1992), 7.

4   Ibid., 2.

5   Leslie Roberts, "Counting on Science at EPA," *Science*, Vol. 249, No. 4969 (Aug. 10, 1990), 616–18.

6   Daniel Wartenberg and Caron Chess, "Risky Business, The Inexact Art of Hazard Assessment," *The Sciences*, Vol. 32, No. 2 (Mar./Apr. 1992), 20.

7   David L. Bazelon, "Risk and Responsibility," *American Bar Association Journal*, Vol. 65 (July 1979), 1067–68.

8   Douglas E. MacLean, "Comparing Values in Environmental Policies: Moral Issues and Moral Arguments," in *Valuing Health Risks, Costs, and Benefits for Environmental Decision Making, Report of a Conference*, eds., P. Bret Hammond and Rob Coppock (Washington, DC: National Academy Press, 1990), 86–88.

9   F. Reed Johnson, Ann Fisher, V. Kerry Smith, and William H. Desvousges, "Informed Choice of Regulated Risk? Lessons From a Study in Radon Risk Communication," *Environment*, Vol. 30, No. 4 (May 1988).

10 A list of references is in ibid., 251, 257.

11 Committee on Risk Perception and Communication, National Research Council, *Improving Risk Communication* (Washington, DC: National Academy Press, 1989), 88.

12 Stephen D. Page, "Indoor Radon: A Case Study in Risk Communication," presented at a conference on *Medicine for the 21st Century: Challenges in Personal and Public Health Promotion*, sponsored by the American Medical Association, the Annenberg Center at Eisenhower, the W. K. Kellogg Foundation, and the Environmental Protection Agency, Palm Springs, CA, Feb. 5–8, 1992.

13 House Subcomm. on Transportation and Hazardous Materials of the Comm. on Energy and Commerce, *Hearing on Radon Awareness and Disclosure*, June 3, 1992 (Washington, DC: Government Printing Office, 1992), 28 (testimony by Michael H. Shapiro).

14 Kristin S. Shrader-Frechette, *Risk and Rationality* (Berkeley, CA: University of California Press, 1991), 5.

15 Letter to Radon Program Review Panel, Environmental Protection Agency, from Anthony V. Nero, Jr., Dec. 10, 1991.

16 William W. Nazaroff and Kevin Teichman, "Indoor Radon: Exploring U.S. Federal Policy for Controlling Human Exposures," *Environmental Science and Technology*, Vol. 24, No. 6 (1990), 776.

17 Interview, June 11, 1992.

18 George F. Kennan, *Realities of American Foreign Policy* (Princeton, NJ: Princeton University Press, 1954), 93.

19 Gabriel A. Almond, *The American People and Foreign Policy* (New York: Frederick A. Praeger, Publisher, 1963), 37, 50–51.

20 William D. Ruckelshaus, "Risk, Science, and Democracy," *Issues in Science and Technology*, Vol. 21, No. 3 (Spring 1985), 22.

21 Committee on the Institutional Means for Assessment of Risks to Public Health, National Research Council, Risk Assessment in the Federal Government: Managing the Process (Washington, DC: National Academy Press, 1983), 3.

22 Ruckelshaus, 28.

23 Mary Douglas and Aaron Wildavsky, *Risk and Culture* (Berkeley, CA: University of California Press, 1983), 71.

24 Shrader-Frechette, 82.

25 Wartenberg and Chess, 21.

26 Committee on the Institutional Means for Assessing Risks to Public Health, 3.

27   Stephen D. Jellinek, "On the Inevitability of Being Wrong," in *Management of Assessed Risk for Carcinogens*, ed., William J. Nicholson (New York: The New York Academy of Sciences, 1981), 43–44.

28   Jeffrey E. Harris, "Environmental Policy Making: Act Now or Wait for More Information?" in Hammond and Coppock, 131.

29   Bazelon, 1069.

30   Steven Wong and Doris Wong, Suzanne Bingham Realties, Inc., v. Gonzalo Mercado and Maria Mercado, Superior Court of New Jersey Law Div.: Bergen County, Docket No. L-40960-87, decided Feb. 7, 1991.

31   William K. Stevens, "Big Increase Expected in Radon Pollution Suits," *New York Times*, 28 Sept. 1986, 54.

32   Interview, July 2, 1991.

33   Nazaroff and Teichman, 780. See also Anthony V. Nero, Jr., "A National Strategy for Indoor Radon," *Issues in Science and Technology*, Vol. 9, No. 1 (Fall 1992), 33–40.

34   In 1992, the EPA indicated that 0.06 percent of all U.S. homes have radon concentrations above 20 picocuries per liter. *Technical Support Document for the 1992 Citizen's Guide to Radon*, E-14.

35   Cited in Giorgio de Santillana, *The Crime of Galileo* (Chicago: University of Chicago Press, 1955), 9.

Appendices

Radioactive Decay Chain of Uranium-238

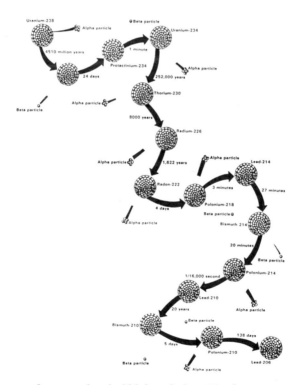

The figure depicts the half-life of the 15 elements and types of ionizing radiation released at each decay step. Radon is the seventh in the series. Reprinted from National Technical Information Service, *Indoor Air Quality Environmental Information Handbook: Radon* (Washington, DC: U.S. Dept. of Energy, Jan. 1986), ch. 1, p. 4.

Letters

FOLLOWING IS THE TEXT OF THREE LETTERS THAT WERE cited in Chapter 1. Written in 1991, they concerned criticism of EPA's radon policies. The first is from Margo Oge, director of the EPA's radon division, to Jane Brody of *The New York Times*. Oge wrote in response to an article by Brody that cited scientists who disagreed with EPA's approach.

The second letter is from Oge to Naomi Harley, a physicist at New York University, who was one of the scientists cited in Brody's article. Oge expressed concern about information that Harley "gave to the press" about her studies of radon. The third letter is from Harley, asking Oge not to misrepresent her publications and work on radon.

UNITED STATES ENVIRONMENTAL PROTECTION AGENCY
WASHINGTON, D.C. 20460

FEB 27 1991

OFFICE OF
AIR AND RADIATION

Jane E. Brody
The New York Times
229 W. 43rd St.
New York, N.Y. 10036

Dear Ms. Brody:

I read your January 8 radon article with great interest as I
have enjoyed your health reporting over the years. After talking
with you on the telephone, I had hoped your article would provide
a broader perspective from the scientific community instead of
the outlying opinions of a few scientists. I was disappointed to
read a rehash of the same old story from the same critics.

It's vexing to see the same three critics deemed as
newsworthy when the vast majority of the scientific community
concurs about the health risks of radon exposure. I was quite
surprised you didn't include any views from the American Lung
Association, the American Medical Association, the National
Academy of Sciences, the World Health Organization, or the
Centers for Disease Control.

I know you have a special interest in cancer risks and
preventative practices, so I'd like to share with you some
additional facts to better explain the magnitude of radon risks.

The **fact** that radon is a confirmed human carcinogen and that
nearly everybody, including Tony Nero and Naomi Harley, agrees it
is linked to thousands of lung cancers deaths each year was lost
in your story! Instead, you reported that some scientists
believe only 100,000 homes are at risk, not 6-8 million. Tony
Nero suggests that only homes above about 20 pCi/L (100,000
homes) should be addressed, not all homes above 4 pCi/L (6-8
million homes).

However, it is critical to note that Nero and Harley do not
disagree with the approximate risks EPA has estimated at 4 pCi/L,
only whether or not these risks deserve the attention of the
government and public. Neither does Nero disagree with the
approximate number of homes at or above 4 pCi/L -- he has
estimated the number to be about 6 million.

2

Rather, Nero asserts that radon risks only become serious when home levels reach 20 picocuries per liter of air (pCi/L) because at 4 pCi/L the risk is similar to day-to-day "routine" risks like car accidents or home fires. Doesn't it sound strange to suggest that the government and private citizens shouldn't be wasting their time on risks like car accidents and home fires? Doesn't nearly everyone agree that the government and private citizens should take steps to reduce these "routine" risks by installing and using seat belts and smoke detectors?

To discount home radon risks at 4 pCi/L is perplexing because citizens voluntarily try to reduce even smaller homes risks by buying items like non-slip bath mats and child-proof latches. Certainly, homeowners are entitled to know radon poses risk even at low levels.

To add some perspective, a home radon level of 4 pCi/L will deliver an effective radiation dose **20 times higher** than the limit allowed under regulations of the Nuclear Regulatory Commission (NRC) for people who live near nuclear power plants. At 4 pCi/L, a person's lifetime risk of developing lung cancer is about 1 in 30 for smokers and about 1 in 300 for people who have never smoked.

Radon's lethal effect is due to its cumulative nature. The risks keep increasing year after year as exposure to radon continues. Individuals with a 4 pCi/L exposure at home are also likely receiving exposures at work, school and elsewhere. While no single home exposure level would prove lethal immediately, hence no trail of death and disease, the cumulative effect of radon may prove lethal years later.

A person can get to a high cumulative radon "dose" by living in a home with a low radon level for many years, or by living in a home with a very high level for a few years. This is why it's impossible to determine an "absolutely" safe radon level.

The public has a right to know that some miners have developed lung cancer at cumulative radon exposures comparable to people living in a home for 50 years with radon levels of 4 pCi/L.

I've enclosed a fact sheet that I hope will help clarify some of the criticisms cited in your story. Clearly, not everything is known about the exact magnitude of risk due to radon exposure. Uncertainties remain just as they do with many health problems; and like virtually all scientific issues there are differing views as to what action the government should take.

3

One thing is for sure. Researchers know more about the effects from radon than most other environmental contaminants and we know enough to take radon very seriously! The public needs the straight story about the health risks of radon -- not debate from a few scientists with their own policy agenda. That story is that thousands of people die each year from lung cancer caused by radon and nearly every scientist agrees -- including Tony Nero and Naomi Harley.

The controversy about radon calls to mind a similar situation with another naturally-occurring carcinogen: nitrates and nitrites. Your nutrition book describes them as: "...among the most potent cancer-causing substances yet discovered, and readily induce several different types of cancers in...animals. Yet no human case of cancer has ever been traced to nitrosamines in foods...."

Your book counseled readers: "A 20-percent cutback in potential exposure to a carcinogen can be significant.... If you want to avoid needless exposure to nitrates and nitrites, start reading labels and restrict your purchases to those products that are prepared without them."

Surely a similar warning about radon is reasonable. Unlike nitrates and nitrites, radon is a known human carcinogen that acts directly to cause cancer, and reducing needless exposure to radon is sensible and simple.

Radon is one of the few serious health risks with a simple and relatively inexpensive solution. I look forward to talking with you in more detail as the documented risks of radon exposure unfold -- just as they did with lung cancer and smoking.

Sincerely yours,

Margo T. Oge

Margo Oge, Director
Radon Division
U.S. Environmental Protection Agency

UNITED STATES ENVIRONMENTAL PROTECTION AGENCY
WASHINGTON, D.C. 20460

OFFICE OF
AIR AND RADIATION

APR   4 1991

Ms. Naomi H. Harley
N.Y. University Department
  Environmental Medicine
550 First Avenue
New York, NY   10016

Dear Naomi:

I was concerned to read about the information you gave to
the press regarding your study of radon and EPA's recommendations
regarding exposure to radon. I do not believe that your remarks
as reported in the press late last week do much to give people an
improved understanding of radon.

I do not believe that EPA has ever recommended that anyone
remediate their home based on basement measurements of radon.
Nor have we suggested that short term measurements in the
basement are representative of personal exposure. I am enclosing
a copy of our "Citizen's Guide to Radon" in case you have not had
a chance to read it. Pages 6-8 provide a clear statement of our
testing recommendations. The guide notes that basement screening
can only be used to indicate that you are likely to not have a
problem if the measurement is under 4 pCi/L. If the measurement
is higher, the guide recommends additional measurements to
determine the long term average throughout the home.

We have also stated in all of our materials that no more
than 10 percent of the homes in America are likely to have annual
average radon concentrations that exceed actions level of 4
pCi/L. We have never the stated that 20 percent of the houses in
the country need remediation. We have said that they needed some
follow-up to determine which ones are truly elevated.

EPA has been accused by some of not being clear enough with
the press to ensure that they do not misunderstand and misreport
our findings and recommendations. Consequently, we are very
cautious to clarify our remarks. I urge you to do the same.
Also, I would appreciate accuracy if you reference EPA's
recommendations.

I believe we all would like the public to better understand
radon, what it means to them, and how they can deal with it
easily and appropriately. Unfortunately, the press is portraying
an ongoing confrontation with Tony Nero, Bernie Cohen, Bill
Mills, and you suggesting that radon is not a health concern, and
EPA, CDC, AMA, ALA and others saying it is a significant health
risk that should be prudently addressed.

I believe that misunderstandings continue regarding your
interpretation of our data and recommendations. Perhaps we have
misunderstood you as well. I suggest that we get together soon
to discuss the issues. Perhaps we can come to a more common view
of the problem and solutions.

Also, I would appreciate receiving a copy of your paper so
that I can understand your methodology and results. I will note
that they are consistent with data that we have collected in
various levels of several thousand homes in different parts of
the country.

Sincerely yours,

Margo T. Oge
Director, Radon Division

NEW YORK UNIVERSITY MEDICAL CENTER
A private university in the public service

Institute of Environmental Medicine

550 FIRST AVENUE, NEW YORK, N.Y. 10016
AREA 212 340-5357

ANTHONY J. LANZA RESEARCH LABORATORIES AT UNIVERSITY VALLEY
LONG MEADOW ROAD, STERLING FOREST, TUXEDO, N.Y.
MAIL AND TELEPHONE ADDRESS: 550 FIRST AVENUE, NEW YORK, N.Y. 10016

April 10, 1991

Ms. Margo T. Oge
Director, Radon Division, ANR-464
Office of Radiation Programs
U.S. Environmental Protection Agency
Washington, DC 20460

Dear Margo:

I received your letter of April 4, 1991 stating that I
"suggest that radon is not a health hazard". I want to point out
to you the fact that I was the first to publish numerical
estimates of lung cancer risk from environmental radon exposure
(Health Physics 40, 307, 1981) and chaired an NCRP Committee
which published the first risk estimates for radon from a
nationally recognized organization.

Because I do not accept the exaggerated risk estimates of
EPA does not indicate that I "suggest that radon is not a health
hazard". I urge you not to misrepresent my publications and
Committee work on radon.

It is also a fact that EPA protocol is to test for radon in
the lowest livable level of the home. Remediation is being based
on such measurement.

You stated in your letter that you were enclosing a copy of
the Citizens' Guide "in case I had not had a chance to read it".
In fact, a copy was not enclosed, however, I have read it quite
carefully beginning with the time I was asked to review and
comment on it when it was in draft form. My comments were
ignored at that time.

I take no position on radon risk, merely report the results
of my research in the matter. The same is true of our recent
study of personal versus home exposure to radon. Our results
showed that basement measurements were meaningless.

Regards.

Sincerely,

Naomi H. Harley, Ph.D.
Research Professor

C

The Environmental Protection Agency/ Advertising Council Messages

THE ENVIRONMENTAL PROTECTION AGENCY/AD COUNcil campaign to encourage people to test their homes for radon began in 1989. The initial campaign was characterized as aggressive by supporters of the material and unduly alarmist by critics. The theme is suggested by excerpts from two pamphlets reproduced here. One begins by saying: "The sooner you fix your Radon problem, the safer you and your family will be." The other says: "Protect your family against Radon…the silent killer."

Critics who call the information misleading cite as an example the caption under a picture of a chest x-ray that says: "Having Radon in your home is like exposing your family to hundreds of chest x-rays yearly." The caption fails to note that radon at some level is present everywhere, and it incorrectly implies that any amount of radon is equivalent to having hundreds of x-rays.

The sooner you fix your Radon problem, the safer you and your family will be.

Take action now.

Radon is the second leading cause of lung cancer in this country. That's why the Environmental Protection Agency (EPA) and the Surgeon General strongly recommend that all homes be tested, and if a problem exists, corrective action be taken as soon as possible.

If your test shows that your home's Radon level is dangerously high, the time to act is now.

How Radon affects you.

Your lung cancer risk from Radon is determined by the amount you're exposed to, and the length of time you're exposed to it. The higher the level, the greater the risk. The chart below gives you an idea of how lifetime exposure to various Radon levels compares with other risks.

RADON RISK EVALUATION CHART

Annual Radon level	If a community of 100 people were exposed to this level:	This risk of dying from lung cancer compares to:
100 pCi/L	About 35 people in the community may die from Radon.	Having 2000 chest x-rays each year
40 pCi/L	About 17 people in the community may die from Radon.	Smoking 2 packs of cigarettes each day
20 pCi/L	About 9 people in the community may die from Radon.	Smoking 1 pack of cigarettes each day
10 pCi/L	About 5 people in the community may die from Radon.	Having 500 chest x-rays each year
4 pCi/L	About 2 people in the community may die from Radon.	Smoking half a pack of cigarettes each day
2 pCi/L	About 1 person in the community may die from Radon.	Having 100 chest x-rays each year

Levels as high as 3500 pCi/L have been found in some homes. The average Radon level outdoors is around .2 pCi/L or less.

The risks shown in this chart are for the general population, including men and women of all ages as well as smokers and non-smokers. Children may be at higher risk.

Interpreting your results.

Because no level of Radon is considered absolutely safe, you should try to reduce Radon levels in your home as much as possible. The average Radon level in homes is about 1.5 picocuries per liter (pCi/L). You should definitely take action to reduce Radon in your home if your average annual level is higher than 4 pCi/L.

In most cases, you can reduce the Radon level in your home to as low as 2 to 4 pCi/L, and sometimes even below 2 pCi/L.

Short- and long-term results should be interpreted differently. If your long-term results are high you should definitely take action to fix your home, as soon as possible.

If your short-term results are high, the best way to determine your annual level is by doing a long-term test of one year. Preliminary research shows that short-term results from tests made during the cooler months generally overestimate annual levels by one to three times.

For example, if your short-term test result is 6 pCi/L, then your annual average level is probably between 2 pCi/L and 6 pCi/L. Or, if your short-term result is 12 pCi/L then your annual average level is probably between 4 pCi/L and 12 pCi/L.

If your short-term test results are low, you may want to test again at some time in the future. This is to make sure that your test was not conducted at a time when Radon levels happened to be much lower than usual.

Fixing your Radon problem.

While in some cases you can treat the problem yourself, you should always consider the use of trained personnel. Trained Radon reduction contractors offer their services in many areas. Call your State Radon Office (see list on back), or your local government to locate one.

You'll find more information on how to reduce Radon levels in your home in EPA's booklet, "Radon Reduction Methods: A Homeowner's Guide," which is also available from your State.

The two most common Radon reduction strategies are:

1. Prevent Radon entry by sealing cracks, sump pump openings, and other areas where Radon can get in.

2. Ventilate the soil surrounding your home so that Radon is drawn away before it can enter your home.

Remember, high levels of Radon are extremely dangerous to you and your family. And if your test results were high, you should take action. Most homes with Radon can usually be fixed for between $200 and $1,500.

So take action now.

Protect your family against Radon... the silent killer.

We all want to protect ourselves and our families. So we keep smoke detectors and first-aid kits in our homes to arm ourselves against disasters.

But there's another hazard that's impossible to see, smell, or touch. Yet it can be found in millions of homes all across America, including your area. It's called Radon.

Radon is a deadly, naturally occurring radioactive gas that causes lung cancer.

Radon can be so deadly that the Environmental Protection Agency and the Surgeon General have strongly recommended that all homes be tested for Radon, except residences above the second floor in multi-level buildings.

Once in your home, Radon can accumulate to dangerously high levels. In fact, Radon is the second leading cause of lung cancer in the United States—after cigarette smoking. As you breathe it in, its decay products become trapped inside your lungs. As these products continue to decay, they release small bursts of energy which can damage lung tissue and lead to lung cancer. It's like exposing your family to hundreds of chest X-rays each year.

However, Radon is easy and inexpensive to detect—and, more importantly, homes with high levels can be fixed.

The risks...how great are they?

Your family's risk of developing lung cancer from Radon depends on the average annual level of Radon in your home, and the amount of time they're exposed to it. Obviously, the longer your exposure, or the higher the level of Radon in your home, the greater the risk.

And that's why it is so important that your home be tested, immediately.

Hopefully, your home won't have a problem. Testing is simple and inexpensive.

And the risk involved if you don't test is great. So the sooner you test your home, the sooner you can take appropriate action.

Testing... short-term vs. long-term.

Radon invades your home from the surrounding soil. In some cases, well water can be a source of Radon.

Once inside, Radon is completely invisible to sight, smell or taste. That's why special detection kits are necessary.

Short-term testing (a few days to several months) is the quickest way to determine if a potential problem exists.

Charcoal canisters, electret ion detectors and alpha track detectors are currently the most common short-term testing devices. Short-term testing should be conducted in the lowest livable area of your home, with the doors and windows shut, during the cooler months of the year.

Long-term testing (up to one year) is the most accurate way to test for Radon. Alpha track detectors and electret ion detectors are the most common long-term testing devices.

Both short-term and long-term testing devices are easy to use and relatively inexpensive.

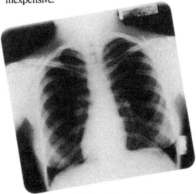

Having Radon in your home is like exposing your family to hundreds of chest x-rays yearly.

Also reproduced here are segments of television advertising that have been part of the campaign.* An initial advertisement showed a family at home. The children and their dog are playing, an older daughter is studying, the mother is on the telephone, and the father is on a couch with his children. A flash image of a skeleton replaces each body, as if its flesh suddenly disintegrated. A voice-over says: "High radon levels will expose your family to as much radiation as having literally hundreds of chest x-rays in one year."

In 1991, the EPA/Ad Council advertising sought humor to inspire people to test for radon. Television spots showed people engaged in silly activities—a woman taking pictures of her bird, a man lining up small statues in his backyard, and members of a fraternal order performing odd rituals. The voice-over was the same for the three advertisements: "What are you doing this weekend that's so important you can't take a little time to test your home for radon? After all it is the second leading cause of lung cancer, and a radon kit is inexpensive."

A third wave of television advertising, planned for 1993, shows a young boy and his dog in gas masks. The boy says the following to the viewing audience: "Everyday a thing called radon gets into people's homes. You can't see it or smell it, but when people keep breathing it they can get lung cancer and die. The thing is, homes with radon problems can be fixed, and it's real easy to find out if you have radon in your home, a lot easier than wearing one of these [gas masks]."

*   Permission granted to reprint material, on pages 221–224, developed by the Advertising Council, Inc., in conjunction with the Environmental Protection Agency's Radon Awareness Program, 1989–1993."

RADON

RADON, THE HEALTH HAZARD IN YOUR HOME
THAT HAS A SIMPLE SOLUTION.

Environmental Protection Agency ♻EPA

"X-RAY" Please discontinue use: October 25, 1990. :30 & :10 versions 30 seconds

ANNCR: (VO) A radioactive gas
has been found

in homes in your area. It's called
Radon.

And it's so deadly,

it's the second leading cause of
lung cancer.

(ELECTRONIC SFX)

High Radon levels will expose
your home to as much radiation

as having literally hundreds of
chest x-rays in one year.

(ELECTRONIC SFX)

(ELECTRONIC SFX)

(ELECTRONIC SFX)

But there is something you can
do about it.

Call to get your Radon test
information.

1-800-SOS-RADON
Ad ♻EPA
Follow your doctor's advice on X-rays

Radon...the health hazard in
your home that has a simple
solution.

Volunteer Agency: TBWA, Advertising, Inc.
Campaign Director: Stephen Kutler, Texaco, Inc.

CNRA-9130/9110

Ad
Council
1089

THE ADVERTISING COUNCIL, INC.

RADON   **Environmental Protection Agency** ♻EPA

Public Service Announcements

Please discontinue use: August 1, 1992

"BIRD LADY" :30 CNRA-1130 (CC) (Also available in :20 length, CNRA-1120 (CC))

WOMAN: Hector, smile for Mommy.

Oh, that's a good Hector.

ANNCR (VO): What are you doing this weekend that's so important

you can't take a little time to test your home for Radon?

After all, it is the second-leading cause of lung cancer and a Radon test kit is inexpensive.

You could pick one up on your way to get more film. WOMAN: (SINGS) FOUR AND TWENTY BLACKBIRDS BAKED IN A PIE.

(BIRD SQUAWKS)

1-800-SOS-RADON.
for more information.
♻EPA

"GNOME" :30 CNRA-1230 (CC) (Also available in :10 length, CNRA-1110(CC)) . And in Spanish: :30 (CNRA-1330)

(MUSIC)

(MUSIC)

ANNCR (VO): What are you doing this weekend that's so important

you can't take a little time to test your home for Radon?

After all, it is the second-leading cause of lung cancer

and a Radon test kit is inexpensive.

You could pick one up on your way to get a few things for the yard.

1-800-SOS-RADON.
for more information.
♻EPA

Volunteer Agency: TBWA, Advertising, Inc.
Campaign Director: Stephen Kutler, Texaco, Inc.

791

THE ADVERTISING COUNCIL, INC.

RADON

Environmental Protection Agency ♻EPA
Public Service Announcements

Please discontinue use: February 15, 1993.

"MOUNTAIN GOAT" :30 CNRA-2130 (Also available in :20 & :10 lengths, CNRA-2120/2110.)

(SFX: GONG, MEN SETTLE DOWN)

LODGE SPOKESGOAT: Brothers

and fellow Mountain Goats...This year's recipient

of the treasured black horn award,

Richard Lynch.

(CHEERS)
C'mon up here, Dick.

ANNCR VO: What are you doing this weekend that's so important

you can't take a little time to test your home for Radon?

After all, it is the second-leading cause of lung cancer

and a Radon test kit is inexpensive.

You could pick one up on your way to the lodge.

1-800-SOS-RADON.
For more information.

(MUSICAL SFX)

Volunteer Agency: TBWA, Advertising, Inc.
Campaign Director: Stephen Kutler, Texaco, Inc.

Ad
Council
192

RADON

THE ADVERTISING COUNCIL, INC.
Environmental Protection Agency ♻EPA
Public Service Announcements

Please discontinue use 8/31/93.

"GAS MASK" :30 CNRA-2430 (Also available in :10 length, CNRA-2410)

CHILD: Every day a thing called Radon

gets into people's homes.

You can't see it or smell it,

but when people keep breathing it,

they can get lung cancer and die.

The thing is— homes with Radon problems can be fixed

and it's real easy to find out

if you have Radon in your home.

A lot easier

than wearing one of these.

(DOG BARKS)

ANNCR VO: Call or test your home today.

Volunteer Agency: TBWA, Advertising, Inc.
Campaign Director: Stephen Kutler, Texaco, Inc.

892

APPENDIX **D**

Differing Perspectives on Radon Policy by the Environmental Protection Agency and the Department of Energy

ON MAY 20, 1992, THE EPA ISSUED A 160-PAGE DOCUMENT to present "the wide range of technical analyses, radon risk communication research, legislative directives, and other information...used to shape the policies that are set forth in the 1992 *A Citizen's Guide to Radon*." Following is a chart from the document's Chapter 6 on "Risk Communication," and excerpts from Chapter 7, which is titled "Rationale for 1992 *Citizen's Guide*."

While the excerpts offer a sense of the rationale behind EPA's policies, a sharply different perspective appeared in a Department of Energy draft memorandum. Also produced on May 20, 1992, the memorandum concerned the Indoor Radon Abatement Reauthorization Act that had been passed by the Senate in March. Beyond critiquing the act, however, the memorandum implicitly addressed the overall EPA approach.

Excerpts from EPA's Technical Support Document for its *1992 Citizen's Guide**

Radon risk communication has provided useful insight into why public apathy exists and suggests directions for overcoming it. Six key findings, listed in Exhibit 6-1, have emerged from the available research that can be applied to the revision of the *Citizen's Guide*.

Exhibit 6-1. Key Risk Communication Findings

1. Be prescriptive as well as informative.
2. Streamline guidelines on testing and mitigation to minimize barriers to public action.
3. Overcome public denial through the use of persuasive appeals such as concern for the family.
4. Provide an appropriate level of radon information, since too much or too little information may result in an undesired effect.
5. Personalize the radon threat with tangible, relevant comparisons to familiar risk.
6. Stress that radon problems can be corrected but do not overstate the ease of fixing them.

How to Communicate Radon Risk Information

As discussed in Chapter 6, extensive risk communication research since the 1986 *Guide* was published has provided useful insight into why the public remains largely apathetic about indoor radon and suggestions for overcoming that apathy. In developing the revised *Guide*, EPA applied six key findings that have emerged from this research: (1) be prescriptive rather than simply informative; (2) streamline guidelines on testing and mitigation so they do not present barriers to public action; (3) overcome public denial through the use of "persuasive appeals" such as concern for the family; (4) provide an appropriate level of radon information, since too much or too little information can result in an undesired effect; (5) personalize the radon threat with tangible, relevant comparisons to familiar risks; and (6) stress that radon problems can be corrected but do not overstate the ease of fixing them.

In updating the *Citizen's Guide*, it was important for EPA to balance the results of this risk communication research with other practical considerations. For example, some of the risk communica-

---

* Radon Division, Office of Radiation Programs, U.S. Environmental Protection Agency, *Technical Support Document for the 1992 Citizen's Guide to Radon* (Washington, DC: U.S. Environmental Protection Agency, May 1992).

tion research suggested that the revised *Guide* should provide only a minimal amount of technical information on radon, since readers can use additional information to create excuses for inaction. This finding, however, had to be balanced against concerns that State offices could be inundated with public requests for more information if the *Guide* did not provide enough detail to answer homeowner's questions. This balancing of the risk communication research with practical limitations led to the following design features with respect to the six key findings listed above:

- The 1992 *Citizen's Guide* was designed to be prescriptive. It provides brief, clear, and easy to follow directions on what to do (e.g., how to test and when to mitigate), rather than simply providing information and allowing readers to come to their own conclusions.

- EPA's Radon Program experience and risk communication research indicate that many people drop out of the testing and mitigation process before they fully comply with the 1986 *Guide's* recommendation to conduct a long-term follow-up measurement prior to reaching a mitigation decision. Based on this finding and EPA's detailed analysis of various testing options, the revised *Guide* attempts to streamline the testing guidelines by: (1) emphasizing the benefits of long-term testing; but also (2) allowing people to choose to conduct either a long- or short-term follow-up test.

- The 1992 *Guide* was designed to function more as a persuasive document than the 1986 version. It utilizes the "protect your family" theme found to be effective in risk communication testing, while avoiding other emotional appeals that might compromise the scientific credibility of the message.

- The revised *Guide* contains a sufficient level of technical information to educate homeowners, while sacrificing as little as possible in terms of clarity, accessibility, and incentive to test.

- EPA designed the revised *Guide* to help personalize the radon threat by providing tangible, relevant comparisons to familiar risks. Specifically, the revised *Guide* provides risk charts for smokers and never smokers that characterize each group's risk as accurately as possible without alienating them or allowing them to deny their risk. The *Guide* also compares radon risk to other risks, such as drunk driving, drowning, fires, airline and car crashes, and violent crimes.

■ The revised *Guide* puts radon mitigation in proper perspective by describing remediation techniques, providing realistic estimates of remediation costs based on EPA's Private Sector Radon Mitigator Survey and research by EPA's Office of Research and Development, and providing examples of other home repairs that are of a comparable cost.

EPA is recommending that the action level of 4 pCi/L established in the 1986 *Guide* be maintained for several reasons. First, lower action levels introduce more uncertainty in the measurement results. Measurement device error increases to approximately 50 percent at 2 pCi/L. This device error in conjunction with the larger fraction of homes (of total homes testing) that have radon levels around 2 pCi/L would result in a threefold increase in false negatives and a twofold increase in false positives over those expected at a 4 pCi/L action level. In addition, as outlined in detail in Chapter 4, the Office of Research and Development's (ORD's) research on mitigation effectiveness and the Office of Radiation Programs' Private Sector Radon Mitigator Survey suggest that elevated levels of radon can be reduced to 4 pCi/L more than 95 percent of the time. Results from the mitigator survey indicate that 2 pCi/L can be achieved about 70 percent of the time, while the ORD research suggests this estimate may be even higher (U.S. EPA/ORD 1989; U.S. EPA/Radon Division 1990a). Reducing the action level to 2 pCi/L, therefore, could result in perhaps as many as 30 percent of homes with elevated levels being unable to achieve the action level.

However, EPA recognizes that mitigation down to lower radon levels may be appropriate because levels below 4 pCi/L still pose a health risk. Furthermore, as mentioned above, mitigation technology available today permits most homes to be reduced to 2 pCi/L or below, and Congress has set a long-term goal that indoor radon levels be no more than outdoor levels, which are typically less than 2 pCi/L. As a result, EPA also closely examined the costs and benefits of selecting an action level of 2 pCi/L and 3 pCi/L.

The results of this cost-effectiveness analysis, detailed in Chapter 5, show that setting the action level at 4 pCi/L would result in a cost of roughly $700,000 per lung cancer death averted (i.e., per life saved). Lowering the action level would incrementally increase this cost to $1,700,000 per life saved if 3 pCi/L was used, and to $2,400,000 per life saved if 2 pCi/L was used instead of 3 pCi/L. All three of the action levels have cost per life saved values that are at the lower end of, or below, the values that the public is willing to pay to save a statistical life, according to EPA's 1983 *Regulatory Impact Analysis Guidelines*.

Based on these findings, EPA believes any of the three action levels considered would provide cost-effective results.

At the selected action level, the Radon Program would be as or more cost-effective than many other government programs for personal safety and environmental protection. EPA also believes the recommended testing protocol combined with an action level of 4 pCi/L in the 1992 *Guide* will be more cost-effective than that recommended in the 1986 *Guide*—$700,000 per life saved now versus the $900,000 per life saved that EPA calculates from program experience based on the 1986 *Guide*.

EPA's decision to keep 4 pCi/L as the action level in the revised *Citizen's Guide* is supported by the cost-effectiveness analysis. However, the revised *Guide* also notes, although not with the same weight as the recommended action level, that homeowners can further reduce their lung cancer risk by mitigating homes that are below 4 pCi/L. As long as the revised *Guide* clearly establishes 4 pCi/L as the recommended action level to avoid confusion and the other problems mentioned above, EPA believes this discussion of reducing radon below the action level helps to fully inform the reader and is justified based on: (1) the health risk involved; (2) the effectiveness of available mitigation technology; (3) cost-effectiveness; and (4) Congressional intent.

In developing the 1992 *Citizen's Guide*, EPA had to balance the findings of its technical analyses on risk, testing accuracy, mitigation technology, and cost-effectiveness with the information it was collecting from its risk communication research. For example, EPA and the scientific community had amassed and analyzed in depth a considerable amount of information on the level and significance of indoor radon risk since the original *Guide* was published in 1986. EPA had to convey that risk in the updated *Guide* with a message that was strong enough to persuade homeowners to act, while being careful not to provide too much detail, which could sacrifice accessibility, or make it too startling, which might compromise scientific credibility and support. Furthermore, although EPA recognized the technical superiority of long-term versus short-term testing after extensive evaluation of the issue, it had to accept the compelling practical limitation that the public at large is more likely to use short-term testing. A lot of "good" testing, after all, will provide greater public health protection than a more limited amount of "perfect" testing. Finally, emphasis on an action level that is achievable by the vast majority of homes is better than recommending a lower action level that pushes the limit of technology. EPA's ultimate objective was to advance a technically well-supported 1992 *Citizen's Guide* that takes a pragmatic step in better communicating radon's risks to the public and promoting broader public action in response to the problem.

EXCERPTS FROM A DEPARTMENT OF ENERGY DRAFT MEMORANDUM*

DEPARTMENT OF ENERGY DRAFT MEMORANDUM (May 20, 1992)
Comparison of Indoor Radon Abatement Act of 1988
vs.
Indoor Radon Abatement Reauthorization Act of 1992*

Sec. 301   National Goal(s)

The Reauthorization maintains the original goal that radon levels within buildings be no higher than of ambient outdoor air.

Adds a second goal--that all homes, schools, and Federal buildings be tested for radon.

Sec. 302   Definitions

Adds: Contracts - For the sale of residential of real property and mortgages, for purposes of requiring radon testing and mitigation in a greater range of U.S. structures. Expands properties from single family houses by this definition to now include all Housing and Urban Development (HUD), Veterans Administration (VA), Farmers Home Administration (FHA), Federal National Mortgage Association, Government National Mortgage Association, and the Federal Home Loan Mortgage Association mortgagees.

Federal Buildings - All Federally-owned or leased spaces.

*The reader should bear in mind that:

EPA crafted this bill's language . . . .

The EPA program will not change if the bill does not pass. Historically, they use radon legislation to provide cover saying "that Congress forces them to do these things" when activities undertaken run counter to scientific knowledge. The program will continue whether the bill passes or not.

The 1988 bill was a costly program building-bill, that did not provide any health benefits. It did provide the EPA program and the radon industry growth opportunities. This bill is a "budget buster" of private sector costs. Important priorities in society have been ignored here, as has smoking as the cause of lung cancer.

The Act does not mention review, consultation, or coordination with radon research programs such as those of DOE. In fact, the DOE is not mentioned anywhere in the Act.

°   U.S. Department of Energy, Draft Memorandum, "Comparison of Indoor Radon Abatement Act of 1988 vs. Indoor Radon Abatement Reauthorization Act of 1992," May 20, 1992.

Sec. 303   Radon Priority Areas (New)

This is an important concept, provides for labeling all U.S. areas as radon "priority" by the Environmental Protection Agency (EPA) administrator if they are likely to exceed a "de mininus amount." (The DOE research estimates approximately 100,000 truly elevated U.S. homes--those above 20 picocuries per liter (pCi/l)--while the EPA, by definition, makes 2/3's of the entire U.S. victim to this Act . . . .   The EPA continues to hinder a . . priority research initiative to facilitate identification of these truly elevated 100,000 homes.

Sec. 304(a)   Inserts - "Centers for Disease Control (CDC) in Consultation with Administrator"

The CDC is included by the EPA to add credibility on the health protection issues. On the other hand, DOE and other active radon research agencies, e.g. the National Institute of Environmental Health Science (NIEHS) and the National Cancer Institute (NCI), disagree with the EPA/CDC position. All involved Federal agencies should participate in policy review and review of the science, if any, supporting EPA policies.

(b)   Citizen's Guide

Reaffirms that EPA radon action levels be "as close to ambient air as possible." No other country in the world does this. This provision does not address the huge private sector cost (over 50 billion dollars), nor the liability issues attendant with lowering guidance below EPA's previously recommended action level of 4 pCi/l.

A major re-measure and re-mitigate effort for all the millions of prior measurements is, thus, implicit. Who is liable? All structures that met 4 pCi/l are now over the new goal. If fact, all structures would require mitigation, because indoor levels exceed ambient.

Sec. 305   Technical Amendments

Deals with development of controversial radon resistant building techniques and their required use in areas designated "elevated" by EPA. At this time, 2/3's of the entire U.S. has been designated as elevated (>2 pCi/l) by EPA. Imposing these techniques will add a significant cost to each new U.S. home built. The multiplier (bringing each new home down to ambient levels) for the U.S. housing industry is not divulged. Cost/benefit is ignored here.

Sec. 306   Technical Assistance

Adds: Develop model state program to disseminate radon information to "tenants" (renters).

Assist states with radon in public water (a major non-problem).

Assist states with adoption and enforcement of EPA new construction standards.

Development of testing guidelines for all multi-story and multi-residence dwellings (represents a major EPA program expansion--large buildings are not a radon problem area. Federal Building Survey confirmed this).

Sec. 306(a) <u>Proficiency Testing</u>

> Important change - EPA national proficiency testing of all radon measurement and mitigation firms is <u>not</u> any longer <u>voluntary</u>. This creates a huge expensive Federal bureaucracy. The radon measurement and mitigation industry wants peers to do certification, <u>not</u> Federal bureaucracy. Radon measurement and mitigation is <u>not</u> brain surgery. Does the Federal Government test and retest termite inspectors? Those who use mandatory state programs can apply for EPA waiver of Federal proficiency test requirement.

Sec. 307   <u>Eligible Activities</u>

> Adds: Public education on radon in <u>private</u> water supplies
>
> Activities to adopt model new construction standards
>
> *Technical and financial assistance to non-profit interest groups to encourage radon testing and mitigation.
>
> *This is a major and troubling provision - under this section EPA will continue to fund millions of dollars of "advocacy" money to the American Lung Association, the National Association of Counties, the radiation control officers, etc. Funds are provided for T-shirts, bumper stickers, media events, scare tactics, legislative support, and advocacy of EPA programs. This allows advocates to be quoted as "supporting EPA" position in radon press releases. These advocates also attack EPA critics.
>
> Targeting outreach to day care facilities (more of above).
>
> Assistance to schools (not new).
>
> Change 3rd year to "each succeeding year"--Federal share will, thus, never end.
>
> Remediation plans for funded schools shall be reviewed by EPA. Why?
>
> Funded states shall investigate consumer complaints that violate EPA proficiency program.

Sec. 308   <u>EPA will Publish School Testing and Remediation Guidelines</u>

> Requires "EPA-certified" person to do all radon measurements and mitigation in schools. This was written in by/for the industry. It is costly. Janitors and building engineers can open charcoal canisters or hang alpha-track detectors. An expensive contractor just multiplies cost.

Sec. 309   <u>EPA Regional Training Centers</u> (Not New)

> Expensive, political, advocacy--these centers established in "politically correct" areas. Well funded, these centers promote radon scares, as well as testing of mitigators and measurers.

Sec. 310   Federal Building Response Plans

>EPA has required all Federal buildings to be surveyed twice now. No survey plan was provided, they were just critiqued after agencies all submitted them. For example, this section covers all military residences, State Department housing, DOE properties, all General Services Administration property. To do this monitoring again, would be truly insane. EPA claims it will "grandfather" this clause and not require retesting. There is no mention of exemptions in the Act. Less than 5 percent of Federal buildings measured at least one room in the building over 4 pCi/l. This section requires that Federal employees cannot measure radon themselves--EPA-approved contractors are required. Another industry-written provision ignores costs and lack of skill required for this task.

Sec. 312   Radon Information

>EPA thrives on the policeman role (given to them in this section) over other Federal agencies.

Sec. 313   Radon Related Information

>Requires information on radon - approved by EPA--be given to buyers in all real estate transfers involving a mortgage.

Sec. 314   Mandatory Radon Proficiency

>Adds mandatory language for EPA radon proficiency program for all measurers and mitigators. EPA can delegate this authority to a State program. Requires costly, nonsensical recordkeeping with EPA having authority to enter and inspect records. This appears to provide Federal authority over a truly non-Federal type of activity.

Sec. 315   Medical Community Outreach

>EPA shall develop and designate radon information to general practitioners, lung cancer doctors, Federal doctors, hospital administrators, and others.

Sec. 316   Federally Owned and Assisted Homes, Schools, and Buildings (new)

>All buildings built with Federal funds must conform to EPA model construction standards.

>HUD must disseminate EPA-approved radon information and protocols to public housing agencies and Indian housing authorities.

>HUD must undertake a radon research program as approved by EPA!

Sec. 317   National Radon Education Campaign

>EPA shall establish a national radon education campaign. This is a frightening provision. What are precedents? Where are U.S. priorities? . . .

Sec. 318   <u>Radon in Workplaces</u> (New)

A major expansion! Federal surveys have shown that it is <u>not</u> a problem, yet this provision making the National Institute for Occupational Safety and Health respond to EPA diverts attention from serious workplace hazards.

Sec. 319   <u>Preemption</u>

This Act does <u>not</u> preempt other laws.

Sec. 320   <u>Enforcement</u>

Provides for civil penalties not to exceed, <u>$25,000 per violation</u>. This breaks new ground. This is unbelievable! EPA is policemen here. Can DOE research findings be seen as violation of EPA intent?

Sec. 321   <u>Citizen Suits</u>

Citizens may bring suit to compel the EPA Administrator to perform his duties under this Act. Will EPA's critics be silenced.

Radon Policies of the American Medical Association and the Health Physics Society

REPORTS BY THE AMERICAN MEDICAL ASSOCIATION largely support the EPA's radon policy, while the position of the Health Physics Society is implicitly critical. Following are excerpts of policy statements by each organization.

The American Medical Association

The Council on Scientific Affairs of the American Medical Association issued two reports on radon, in 1987 and 1991. Both were adopted by the association as official policy. Following are the abstracts of each and their concluding recommendations.

From the 1987 Report of the Council on Scientific Affairs:

> Radon 222 and its radioactive decay products can enter buildings and, through inhalation, expose the inhabitants' pulmonary tissues to ionizing radiation. Studies of radon levels in the United States indicate that variations of 100-fold or greater exist among private dwellings. In one region, 55 percent of homes had levels exceeding 4 pCi/L, which is the guidance level recommended by the US Environmental Protection Agency. Ventilation and tightness of construction are important determinants of radon levels. In some instances, fans or heat exchangers can reduce excessive concentrations, but in others more elaborate remedial measures may be required. Physicians may obtain information about radon through Environmental Protection Agency regional offices and state radiation control programs. The risk of radiogenic cancer is believed to increase with exposure to ionizing radiation. According to some estimates, concentrations of radon decay products in US homes could be responsible for several thousand cases of lung cancer per

year. Studies of radon levels in representative buildings and guidelines are needed to ensure safe, effective, and cost-effective countermeasures. Architects, contractors, designers, building code administrators, health physicists, and biomedical investigators can help with solutions....

The Council on Scientific Affairs recommends the following:

1. The American Medical Association should assume a leadership role in educating physicians, others of the health care community, and the public concerning the significance of radon levels in homes and other buildings and the possible health effects of those levels.

2. The American Medical Association should take the lead in arranging with the US EPA, the US Department of Energy, the National Institute of Environmental Sciences, and other agencies to sponsor a conference on the effects of radon and their mitigation.

3. Studies of ambient radon levels should be conducted in representative homes, commercial buildings, and enclosed workplaces of the nation.

REPORT OF THE COUNCIL ON SCIENTIFIC AFFAIRS, AMERICAN MEDICAL ASSOCIATION, "RADON IN HOMES, *JOURNAL OF THE AMERICAN MEDICAL ASSOCIATION,* VOL. 258, NO. 5 (AUG. 7, 1987), 668-72.

From the 1991 Report of the Council on Scientific Affairs:

The consensus of scientists is that exposure to radon is hazardous, but disagreement exists about the effects of lower radon concentrations. Studies of underground miners have indicated that the risk of lung cancer increases in proportion to the intensity and duration of exposure to radon, and a recent authoritative report (BEIR IV) has concluded that estimates based on those studies are appropriate for estimating risks for occupants of homes. The BEIR IV report concluded that smoking cigarettes incre .ses the risk of lung cancer associated with radon. Average radon levels in US homes range from 1.5 to 4 pCi/L, depending on the circumstances of measurement. Few studies have investigated health outcomes in occupants of homes with high radon levels. In advising patients about reducing the risks associated with radon, physicians should consider the costs, as well as the benefits, of remedial actions, and they should emphasize that, by far, the best way to avoid lung cancer is to stop smoking....

The AMA Council on Scientific Affairs makes the following recommendations:

1. The AMA should continue its surveillance of the growing understanding of the health risks of exposure to radon and contribute to this understanding wherever possible.

2. Physicians should continue to increase their knowledge about radon and its health effects and advise patients and the public in their communities on how to make intelligent decisions and take responsible actions on this issue.

3. Physicians, when discussing the prevention of lung cancer,

should place the greatest emphasis on the need to stop smoking. Measures to decrease radon exposure should be encouraged, if appropriate, but placed in proper perspective because smoking is a more significant cause of lung cancer.

4. The AMA should emphasize the need for more definitive data concerning the magnitude of the lung cancer risk from radon exposure and encourage the generation of these data as a needed public health measure.

5. The AMA should continue its efforts to help physicians understand the health risks associated with radon exposure and communicate this understanding to patients and the public.

REPORT OF THE COUNCIL ON SCIENTIFIC AFFAIRS, AMERICAN MEDICAL ASSOCIATION, "HEALTH EFFECTS OF RADON EXPOSURE," *ARCHIVES OF INTERNAL MEDICINE,* VOL. 151 (APR. 1991), 674-77.

Health Physics Society Position Statement*

Perspectives and Recommendations on Indoor Radon
October 1990

Radon is a colorless and odorless radioactive gas that is and always has been a natural component of the air we breathe. Radon is produced by the radioactive decay of radium, a naturally occurring radioactive element that is found in trace amounts in all soils as well as building materials, plants, animals, and the human body. Although scientists have been aware of radon for many years, it was not until recently that it was realized that the largest radiation exposures received by most individuals come from natural sources of radiation, primarily radon and its radioactive decay products. This new understanding of the role of radon has led to anxiety over radiation exposures among members of the general public and considerable and often inaccurate statements in the media.

The Health Physics Society encourages public understanding of the potential risks from radon, and recommends that exposure to radon and its radioactive progeny be minimized in accordance with practical considerations, taking into account applicable technical, economic, and societal factors. To assist the public, health officials, educators, and the media, the Society offers the following seven observations and recommendations on the topics of radon exposures, risks to health, and national abatement programs.

---

* *Newsletter,* Health Physics Society, Vol. 19, No. 1 (Jan. 1991). Reprinted with permission.

1. Provide the public with realistic expectations.

The assumption that radon contributes significantly to the total number of lung cancers may be reasonable, but cigarette smoking is still, by far, the major cause of lung cancer and will completely dominate the incidence of this disease for the foreseeable future. Only after many years would a successful radon abatement program begun today be likely to reduce the number of lung cancers, and then only by a very small percentage of the total.

2. Base priorities on the likelihood of exposure.

The Environmental Protection Agency should review its emphasis on the use of 4 picocuries of radon per liter of air (pCi/L) as an action level. Rather the EPA should emphasize the prompt identification of indoor occupied areas with very high radon concentrations (i.e., tens of pCi/L and greater) as candidates for prompt mitigation. The EPA and other responsible public health officials should encourage appropriate radon testing to identify suitable candidate buildings.

3. Inform homeowners of the benefits of radon reduction.

The EPA and other public health agencies should provide realistic information on the potential benefits of radon reduction in homes, based on the actual lifetime radon exposure that will be avoided. The potential synergistic effects of cigarette smoking and other air pollutants should be included in this information. Homeowners should be informed that the benefits to individuals from radon reduction are not necessarily based on the initial radon concentration, but rather the degree of reduction achievable and other factors related to home occupancy and smoking habits.

4. Inform elected officials of the benefits of radon reduction in schools and other public buildings.

Elected officials should be provided with realistic information on the benefits, costs, and practicality of radon reduction in schools and other public buildings. Evaluation of potential benefits should include consideration of the number of people exposed and the magnitude and expected long-term effectiveness of exposure reduction. The Health Physics Society encourages elected and other officials to make decisions regarding abatement based on measurements that are representative of conditions of normal occupancy.

5. Appeal to reason rather than emotions.

Education of the public should be based on reason, rather than emotion. In particular, the Society condemns the use of fear and other emotional scare tactics to overcome apathy or encourage action on the part of homeowners and other members of the public. Enhancement of the already present and lamentable radiation phobia in the public may easily have unexpected and undesirable repercussion such as reluctance to undergo valuable and potentially beneficial medical procedures due to an irrational fear of radiation effects.

6. Redirect the National Radon Abatement Program.

The EPA should place its major emphases for radon abatement in three areas: (1) radon-resistant design for new construction; (2) identification and mitigation of the highest indoor radon exposures, i.e., homes, schools, hospitals, and other public buildings with radon concentrations of tens of pCi/L or greater; and (3) assistance to the real estate industry and mortgage lenders in developing improved programs of radon testing and abatement as necessary at the time of property transfer.

7. Encourage and support additional research.

Although we know a great deal about radon and its potential effects on health, there is still much we do not know and could benefit from learning. The EPA and other governmental agencies concerned with radiological health should encourage and fund additional research by competent qualified scientists to improve our understanding of the risks of radiation and the means to mitigate those risks.

Index